Industrial Robots
and
Robotics

Industrial Robots and Robotics

Edward Kafrissen

Director, Robotics Research Laboratory
New York Institute of Technology

Mark Stephans

Reston Publishing Company, Inc.
Reston, Virginia
A Prentice-Hall Company

Library of Congress Cataloging in Publication Data

Kafrissen, Edward.
  Industrial robots and robotics.

  Bibliography: p.
  Includes index.
  1. Robots, Industrial. 2. Robotics. I. Stephans,
Mark. II. Title.
TS191.8.K34 1984   629.8'92   83–17723
  ISBN   0–8359–3071–8

Copyright © 1984 by Reston Publishing Company, Inc.
A Prentice-Hall Company
Reston, Virginia 22090

Editorial/production supervision and interior design by
Ann L. Mohan

10 9 8 7 6 5 4 3

PRINTED IN THE UNITED STATES OF AMERICA

To my father, Abraham
and my friends, Alicia and Phillip.
Edward Kafrissen

and

In memory of my friend, Louis Fridovich.
This book is also dedicated to
his wife, Sylvia, and
his children, David, Bernard, and Irwin.

Mark Stephans

Contents

3   The Servo Control System of a Robot   63

4   The Actuators of a Robot   89

Appendixes

Preface

This text has been written for an introductory course in robots and robotics to be used in community or senior colleges as an elective course.

Chapter 1 gives an overview of industrial robots for today and tomorrow. Chapter 2 treats the concept of computers and microprocessors used in robot and robotic work. Servo control systems of robot comprise Chapter 3. How the actuator of a robot converts hydraulic, electrical, and pneumatic energy to effect the motion of the robot is explained in Chapter 4. Chapter 5 takes up robotic transducers including resolvers, encoders, and other sensors, followed by data acquisition and conversion systems. Chapter 6 includes harmonic drives, ball bearing screws, shock absorbers, and other systems. Chapter 7 takes up the important topic of robotic vision. Chapter 8 includes practical industrial robots, specifications, programming, and applications. Chapter 9 emphasizes robot software including program language for robots. Chapter 10 concentrates on setting up robotic work cells. Chapter 11 highlights the justification and implications of economics and growth in robotics. The appendixes include robot terminology, robot manufacturers, Air Force and Army robotic programs, a general bibliography, and other information useful in robotic work and study.

This text can also be used as a reference for anyone interested in learning about robots—engineers, manufacturers, and hobbyists. Each chapter ends with problems and references.

The authors have tried to present the material in a logical and consistent fashion. Since this is a relatively new field, many concepts that are presented today may be changing tomorrow.

There will be many jobs in the years to come in robots and robotics. You may be a member of one of the groups that make themselves available for work in the robot technology era whether you are in the electronic, computer, medical, mechanical, or electromechanical field, to name a few areas of interest.

In *The Britannic Review of Development of Engineering Education,* published by Encyclopaedia Britannica, Inc., Lord Kelvin had this to say about measurement and instrumentation:

I often say that when you can measure what you are speaking about, and express it in numbers, you know something about it. But when you cannot measure it, when you cannot express it in numbers, your knowledge is of a meager and unsatisfactory kind. It may be the beginning of knowledge, but you have scarcely, in your thoughts, advanced to the stage of science, whatever the matter may be.

(From Volume 1, page 123, © 1970 by Encyclopaedia Brittanica, Inc.)

People who are interested in robotics will have to learn by experience and apply this technology to better themselves in this thriving new field.

The authors wish to thank Mr. David Ungerer, president; Mr. David Dusthimer; Mr. Greg Michael; Mrs. Linda MacInnes, editorial assistant; Mrs. Ann Mohan, production editor; and the entire crew of Reston Publishing Company. Without their help this project would never have gotten off the ground floor.

The authors would also like to thank Mrs. Betty Johnson, who typed the original manuscript, and express condolence on the accidental death of her 22-year-old son, Eric Johnson.

Edward Kafrissen
Mark Stephans

Acknowledgments

This is to acknowledge the manufacturers and societies who have con-
tributed to this text. Special thanks must go to Mr. John Lawson, President of
Feedback, Inc., who provided useful information in the development of this
text. Also, the authors wish to thank Dr. John W. Hill, Vice-President of
Microbot, Inc. for his contribution.

Industrial Robots and Robotics

1

Industrial Robots Today and Tomorrow: An Overview

1.1   Introduction

In the early 1960s, George Devol and Unimation Incorporated introduced the first industrial robot. The basic concept was to build a machine that was flexible enough to do a variety of jobs automatically, a device that could be easily taught or programmed so that, if the part or process changed, the robot could adapt to its new job without expensive retooling, as was the case with hard automation. It was this mating of a computer to a flexible manipulator that helped to open the door to new methods of manufacturing.

James S. Albus, in a recent book on the effect of computers and robots, wrote

---

Text and diagrams for Chapter 1 are in part based on material supplied by Jerome W. Saverino, Vice-President, Marketing and Applications, LTI Technology, Suite 701, 2701 Toledo St., Torrence, Calif. 90503.

From Jerry W. Saveriano, "Industrial Robots Today and Tomorrow," *Robotics Age Magazine,* Summer 1980, Vol. 2, No. 2. Reprinted with permission of Robotics Age, Inc., 174 Concord Street, Peterborough, N.H. 03458.

The human race is now poised on the brink of a new industrial revolution which will at least equal, if not far exceed, the first Industrial Revolution in its impact on mankind. The first Industrial Revolution was based on the substitution of mechanical energy for muscle power. The next industrial revolution will be based on the substitution of electronic computers for the human brain in the control of machines and industrial processes.[1]

Recent advances in microprocessor technology, the increasing need to improve productivity, and the long-range goal of computer-controlled, unmanned manufacturing plants have provided new impetus for industry to apply robots in manufacturing. *Fortune Magazine,* in its December 17, 1979, issue, stated, "The number of industrial robots in use in the United States has more than doubled in the past three years. The industry's sales have been growing at an annual rate of thirty-five percent."[2] Ten years from now, sales in the United States are expected to total over $5 billion.

In 1982, there was on the order of 6000 robots and robotic devices in use in the American industry. By 1985, the projected number increases to about 20,000 in use in the United States. In 1995, the automated manufacturer and the plastic processors will use around 200,000 industrial robots and robotic devices.

Estimates of the total number of industrial robots and robotic devices in use throughout the world in 1982 include 6000 in the United States, 5000 in Europe, and approximately 30,000 in Japan. Thus approximately 41,000 industrial robots are already in use throughout the world.

Included in the count in Japan are simple robots similar to the Seiko Model 700 shown in Figure 1.1. These types of robot devices have been used in Japan for almost 40 years.

In a paper given to the Japan Industrial Robot Association, the executive director, Kanji Yonemoto, gave this as the general view of robot use in Japan:

By virtue of their functional characteristics enabling versatile and flexible motions, industrial robots help achieve automation of the small-batch of mixed-line production, thereby contributing greatly to the improvement of productivity.

Utilization of industrial robots brings about a change of the production system from the "man–machine system" to "man–robot–machine system." This change brings socioeconomic impacts such as the improvement of working environments and the humanization of working life, by helping promote industrial safety and advance labor quality, as well as increase returns on investment, stability of product quality, and improvement of production management.[3]

There is little doubt that computers and robots will significantly alter manufacturing in the near future. To gain a better understanding of this exciting and new technology, it is best to start with some fundamentals of robotics and then advance to robot systems.

The Robert robot as shown in Figure 1.2 is a simple two-axis ROBOT designed to teach the principles of Robots. Robert has the following features:

OPERATING RANGE

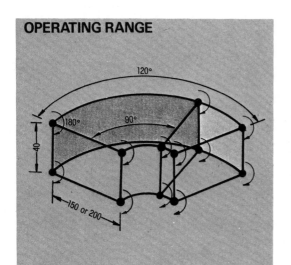

TECHNICAL DATA

Vertical 垂 直	40 mm (40 mm/0.5 sec)
Horizontal 水 平	150$^{\pm 2.5}$ or 200$^{\pm 2.5}$ mm (150 mm/0.5 sec)
Swing 旋 回	90°$^{\pm 2°}$ or 120°$^{\pm 2°}$ (clockwise or counterclockwise) (90°/0.5 sec) (正 逆 万 向)
Repeatability 再 現 精 度	±0.025 mm
Payload (incl. gripper) グリッパ含可載重量	500 g (1000g at lower speed) (スピードが遅い時1,000g)
Dimensions 寸 法	280 × 220 × 230 mm
Weight 重 量	20 kg
Drive 駆 動	Air pressure normal use 4 kg/cm² (常用空気圧)
Option オプション	Horizontal axis 250 mm long 水 平 軸 **Gripper** (open + close) (W: 300g) グリッパ (開+閉)

DIMENSIONS

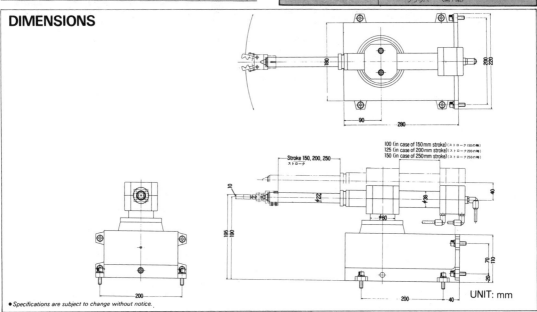

• Specifications are subject to change without notice.

UNIT: mm

FIGURE 1.1. Specifications for Seiko Model 700, a four-axis air-operated simple robot arm. Courtesy of Seiko Instruments, U.S.A., Inc., Torrence, Calif. [See next page for photo.]

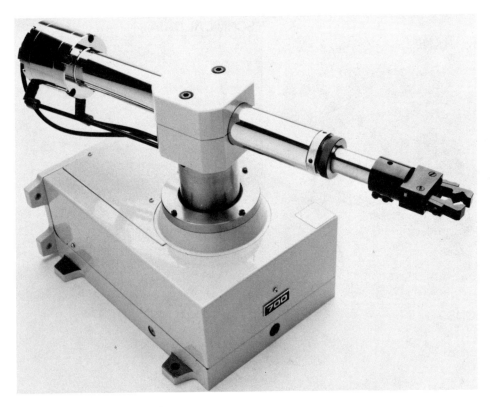

FIGURE 1.1. [Continued].

1. Robert had been designed as a suitable study case for an introduction to robotics or, more simply, to illustrate the method of controlling stepper motors, solenoids, and the like, by a suitably programmed microcomputer.
2. Robert is supplied complete with a magazine and a selection of suitable resistors together with a "Personality Pack," which contains:
   a. Software cassette
   b. Printout of program using BASIC
   c. Flow chart of program
   d. Interconnection lead between Robert and your microcomputer or microprocessor
3. Robert has both linear and rotary movement with an arm-mounted gripper. Movement is actuated by stepper motors with 48 steps/revolution resolution, which through gearing allows quite accurate positioning of the gripper.

FIGURE 1.2. Robert the intelligent robot. Courtesy of Feedback Inc., Berkeley Heights, N.J.

1.2   Basic Robot Elements

The definition of a robot differs among experts all around the world. According to the Robot Institute of America, a robot is a "reprogrammable multifunctional manipulator." Many robot watchers reckon that the word "reprogrammable" in this definition means that a mechanical arm must be computer-controlled to qualify as a robot. The Japanese usually include "fixed-sequence robots" in their statistics. The mechanical arm of these machines is not controlled by a computer, but by a series of electromechanical switches. The following sections, 1.2.1 through 1.2.3, cover the three basic components of an industrial robot.

1.2.1   Controller[a]

The controller performs as many as three functions: (1) storing, sequencing and, for more sophisticated robots, positioning of data in memory; (2) initiating and, for some robots, stopping the motions of the manipulator (in a speci-

---

[a]Section 1.2.1 is used courtesy of L. F. Rothschild, Unterberg, Tobin, New York City.

fied sequence) at given points; and (3) interacting with the external environment. The controller has two basic components, the hardware (the computers) and the software (the implicit operating and task programs). These two programs access an input program (data describing a sequence of movements for the task at hand), process the data, and trigger electrical signals that prompt mechanical actions to achieve specific motions required to perform a task. The controller interacts with the actuators' control valves and circuits to either prompt movement of the manipulator by creating a motion control loop (*initiating* motion) or, for more advanced systems, governs motion by creating continuous control and feedback systems (*controlled* motion). The main computer is either an integral part of the manipulator or is stored in a separate cabinet (i.e., is either onboard or outboard). Feedback devices for the joints are always housed within the robot's manipulator (discussed in Section 1.2.2).

The controller is mechanical (cams, drums), pneumatic, or electronic device. The latter is most frequently used today. Electronic controllers are either numerical controllers or, more commonly, flexible digital computers. The computer records (and later plays back) prescribed directions for sequential manipulator movements for executing a task, the input program. These data are processed, converted into electrical signals, and sent to each axis's actuator to move that element. Low-level technology robots use step-sequencer-type controls to access each step of the program, while higher-level technology robots use more advanced computing systems, such as microprocessors, to monitor the coordination of the axes' motions in executing the program. More sophisticated feedback systems receive and interpret information from the controller. This requires (1) the processing and recording of each motion of the task to create data for a program, (2) computing the amount and direction of movement for each axis (to effect the motions required so that the robot can carry out the directives inherent to the program), and (3) transforming these computations into electrical signals to prompt *controlled* actions. After each motion, internal sensors register the positions of the links and joints within the manipulator and/or external sensors track the position of the end-effector (or machinery with which the robot may be interacting). This information is transmitted back to the computer, creating *two-way* communication between the controller and the joints, between the controller and the end-effector, or between the controller and any ancillary devices, *allowing for continuous monitoring of all robot motion.*

To initiate the required motions to perform a task and, for the more sophisticated robots, to maintain control of the manipulator's elements (after a movement is initiated), the computer must be programmed. A program, effectively, sets a series of target positions through which the end-effector must pass to perform a given task. Programming is accomplished by (1) manually installing a program (e.g., the pneumatic fluid stop SEIKO arm), (2) physically leading the robot through a sequence of motions (e.g., the painting robots), or

(3) using machine input controls, via a teach pendant or a computer terminal (e.g., the assembly robots such as the PUMA or MAKER).

1.2.2   Manipulator[b]

The manipulator is typically cylindrical or rectangular in shape. The base is usually stationary, but in some work setups the robot can be moved to different work stations via a track or rail. From the base extends an appendage. This consists of an arm, either straight or jointed (i.e., articulated), to which is attached the wrist and a wrist flange; the latter allows for rotation of the end-effector. The appendage is internally powered and moves the end-effector to achieve a given target position within the work space/envelope. The robot shown in Figure 1.3 is a jointed spherical, servo-controlled, continuous path arm with five revolute axes.

Within the manipulator are the mechanical parts—the links, powered joints, and elements—that execute the robot's movements. When the robot's motors are hydraulically or pneumatically powered, specialized high-precision valves (servovalves) control the flow of oil or air. When the motors are electrically powered, special circuits control the flow of current. These motors are small and, whether oil, air, or electrically powered, are referred to as *actuators*. Each joint of the robot has up to three axes (i.e., three per plane), and each axis has its own actuator. The actuators are located at the robot's joints and are coupled to mechanical linkages that power and direct the motion of the connected element and, in so doing, move the manipulator. This coupling may be direct or indirect; in the latter case, gears, chains, or ball screws are used.

1.2.3   Tooling

The hand or gripper, sometimes called the end-effector, can be a mechanical, vacuum, or magnetic device for a part handling. (See Chapter 6 for a detailed discussion of grippers.) At present, most of these devices are only two position (i.e., open/closed, on/off).

Figure 1.4a and b shows two simple hands used for grabbing cylindrical parts. The fingers are easily changed to accommodate different part configurations. Figure 1.4c shows a gripper that holds the workpiece, either by the I.D. or O.D., to facilitate loading or transfer. Figure 1.4d is a double hand that also grabs the I.D. or O.D. The dual hand is advantageous because it enables the robot to unload a machine tool with one hand while holding the unprocessed part in the other. This allows the robot to reload the machine without moving

---

[b]Section 1.2.2 is used courtesy of L. F. Rothschild, Unterberg, Tobin, New York City.

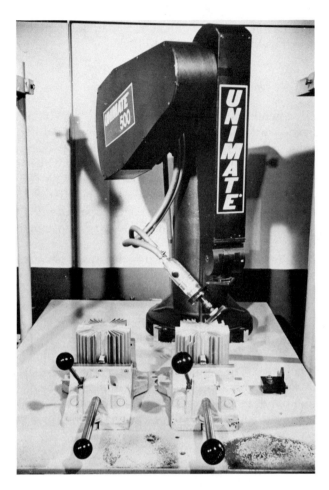

FIGURE 1.3. PUMA 500™, a five-axis electric-powered robot doing deburring. Courtesy of Unimation® Inc., Danbury, Conn.

back to a pickup point to get the next part. Figure 1.4e is a special gripper used in the die-cast machine work cell shown in Figure 1.11. The lower part of the hand loads the two inserts into the die cavity. After the die-cast machine cycles, the upper part of the hand unloads the cavity.

Robots are not only used for material handling; they are also very effective tool handlers. Figure 1.4f, g, h and i shows some common tools that can be manipulated by industrial robots. Figure 1.4j is a quick disconnect coupling at the end of the robot's wrist so that it can change tools automatically. The variety of tools and grippers that can be adapted for robot use is unlimited.

It is, however, the gripper of today's industrial robot that is one of the most limiting factors in universal robot utilization due to the lack of hand

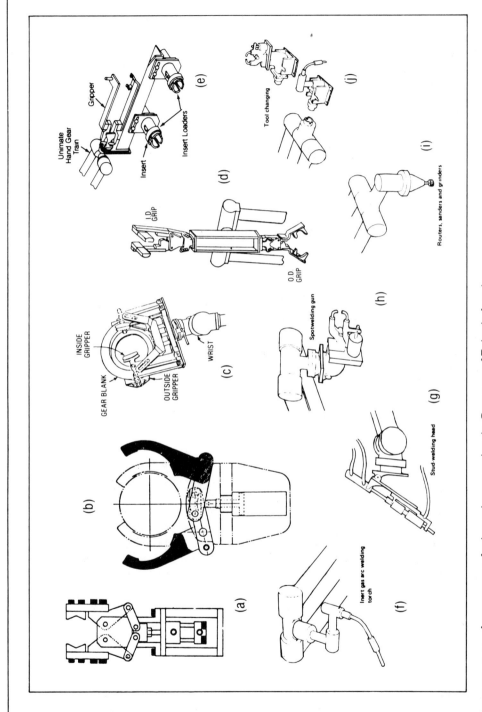

FIGURE 1.4. Assortment of robot grippers and tools. Courtesy of Robotic Age, Inc., Peterborough, N.H.

9

programmability. It is the weakest link of the robot's components. Extensive research and development are being done to produce a gripper that can handle a wide assortment of part configurations.

One of the greatest lessons in the study of artificial intelligence and robotics is the recognition of the marvels of the magnificent human machine.

1.3   Types of Industrial Robots

As varied as the definition of "robot" are the many ways to classify different types of robots. Under the broad classification of industrial robots, categories may be based on the kind of work the robot is assigned to do, such as spray painting, welding, assembly, and material handling.

1.3.1   Simple Robots

Simple robots are also called *pick and place* devices and *limited sequence* manipulators. Simple robots are the most underrated and underutilized robots in the United States. These low-cost, easy-to-maintain, fast, and accurate devices can dramatically increase productivity in medium- and long-run production industries.

Simple robots like the one shown in Figure 1.1 are usually very limited in the amount of information that can be stored in their memory. Generally, only sequence and time are used in their programs, although some branching and subroutines are possible with today's low-cost, increasingly powerful programmable controllers.

Normally, these devices are restricted to three or four nonservo degrees of freedom. Mechanical stops are used on each axis to set the amount of travel; this is usually only two positions (i.e., up/down, right/left, in/out). Because they are very limited in the number of moves available to the manipulator, simple robots are very dependent on support equipment such as bowl feeders and part presenters. A general rule of thumb in robotics is that the higher the intelligence of the controller and the greater the programmability and number of moves of the manipulator and tooling, the less dependent the robot will be on support equipment.

Simple robots are usually air operated, accurate to ± 0.001 in. or better, and can operate as fast as 6 hertz (Hz) or cycles per second.

1.3.2   Medium-Technology Robots

Medium technology robots have a greater memory capacity and are easier to teach than simple robots. Examples of medium technology robot work cells are given in Figure 1.5, which shows machine loading with the Unimate 2000's robots and Puma robots on a conveyor line, palletizing. Work cells include the environment where the robots perform. Such robots have four to six degrees of freedom and are servo-controlled in most of their axes of movement. The robot shown in Figure 1.6 is also a good example of a medium-technology robot. It has a 256-step memory and is programmable in its three major axes (i.e., waist rotation, radial traverse, and vertical traverse). The wrist bend and rotation are nonservo and set with mechanical stops as with the simple robot. Figure 1.7 is another example of a medium-technology robot.

Medium-technology robots are usually used in single machine load/unload types of jobs and are not capable of continuous-path operations required for welding and spray painting applications. There are many jobs in manufacturing today that could be automated by using medium-technology robots. Such units generally have a repeatability of ± 0.050 in.

1.3.3 Sophisticated Industrial Robots

Sophisticated industrial robots are considered by many to be the superstars of the robot family and are at the leading edge of manufacturing technology. These robots possess highly flexible and programmable manipulators and utilize controllers that exemplify the highest level of artificial intelligence used in industrial automation. Such controllers can be interfaced with sophisticated sensory and inspection devices and also enable the robot to be taught even the most complex of jobs with relative ease. The sophisticated industrial robot has the capability of being integrated into a myriad of computer-controlled work cells and systems.

Today's sophisticated robots, such as the PUMA arm shown in Figure 1.8 and Cincinnati Milacron's T^3 robot shown in Figure 1.9, are truly amazing tools. These machines have given computers the ability to move in space and can be thought of as the physical extension of the computer in the real world.

Sophisticated industrial robots have a large onboard memory, capable of containing multiple programs and the ability to change programs automatically, depending upon the requirements of the work cell or system in which they are working. These machines are easily programmed by a teach pendant, terminal keyboard, or off-line programming, or any combination of the three. Limited voice control is becoming available. High-level robot programming languages and software are being used. Sophisticated robots' controllers are usually micro- or minicomputers; program storage can be on any number of available media. The manipulators have five or more degrees of freedom and are fully programmable in all axes: these manipulators can operate either point to point or continous path. Small sophisticated robot arms such as the PUMA have repeatability of ± 0.004 in. carrying loads of a few pounds.

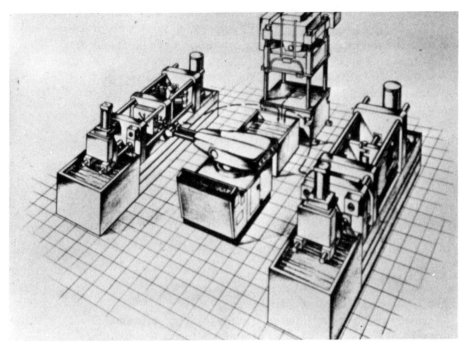

A] Machine loading with the Unimate 2000s.

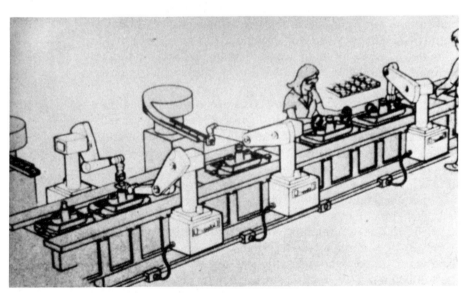

B] Puma robots on a conveyor line, palletizing.

FIGURE 1.5. Medium-technology robot work cells. Courtesy of Unimation® Inc., Danbury, Conn.

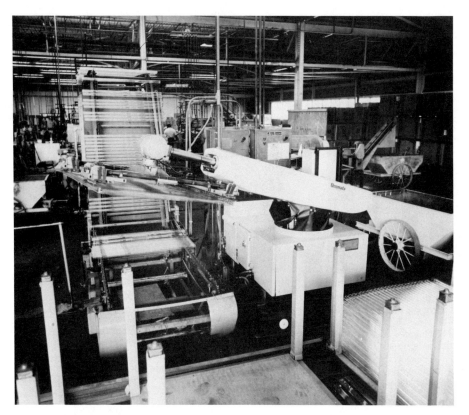

FIGURE 1.6. Unimatic Model 1000, a five-axis, hydraulic-operated, medium-technology robot performing glass handling. Courtesy of Unimation® Inc., Danbury, Conn.

It is important to realize that these classifications are relative and in fact represent a broad continuum of capabilities. As the industry progresses, robots will change classification: today's "sophisticated" robot will be tomorrow's "medium-technology" robot.

When a robot is classified by function, it falls into four main categories:

1. A pick-and-place (PNP) robot operates to pick up an object and move it to another location. Typical applications include machine loading and unloading, palletizing, stacking, and general materials handling tasks.
2. A point-to-point (PTP) robot moves from point to point and can move internally hundred of points in sequence. At each point, the robot performs a function, such as spot welding, gluing, drilling, or deburring.
3. A continuous-path (CP) robot moves from point to point, but the path is critical. Typical applications include paint spraying, seam welding, cutting, and inspection.

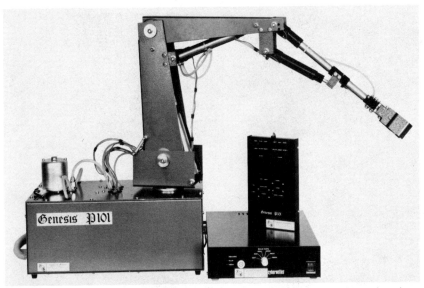

The left picture shows the Genesis P101 robot. The right hand top picture shows the microprocessor box that controls the robot, and the bottom box provides entry commands via the teach
pendant.
      The Genesis P101 is an industrial robot not necessarily used on the production line but for
proofing applications and software. Courtesy of Feedback, Inc.

FIGURE 1.7. The Genesis P101, a medium-sized robot and accessories. Courtesy
of Feedback, Inc., Berkeley Heights, N.J.

4. Assembly robots combine the path control of CP robots in precision
   work of machine tools.

1.4   Examples of Current Applications

1.4.1   Press Loading

Figure 1.10 shows the layout of a simple robot loading a press. The part is fed
to the robot by the part feeding equipment. The escapement device presents
the robot grip with the part. A proximity switch of some kind would be built
into the escapement device, signaling the robot that a part was indeed present.
The robot would grasp the part and pull it from the escapement device; as it
rotates toward the press, it flips its grip 180°, inverting the part. All the moves
occur synchronously. The arm reaches its left rotary stop, then reaches in and
lowers the part into the die. When the arm is withdrawn from the press, the
robot's internal sensors notify its controller that the arm is clear of the press;

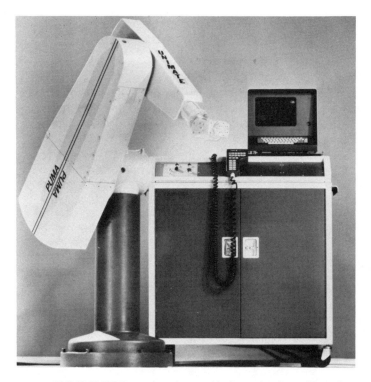

FIGURE 1.8. PUMA™ 760, a six-axis, sophisticated robot. Also shown are the VAL™ computer control, teach pendant, and computer terminal. Courtesy of Unimation® Inc., Danbury, Conn.

the controller would then cycle the press. The part is either dropped through the die or blown off to a chute behind the press. The blow off and the part ejection are monitored by the robot's controller. As this is occurring, the robot swings back to the part feeder and is ready to begin the next cycle.

If a part jams in the feeding device or is not ejected from the die, or if any other sequence did not function as programmed, the controller would instantaneously shut down the work cell and signal the setup man via an annunciator device.

This simple operation could be run three shifts a day, seven days a week, and only require an operator to load the hopper with new parts and to remotely monitor its running. A dramatic labor savings and production increase is obvious in this kind of application, in addition to the ultimate elimination of boring, repetitive, and potentially unsafe jobs, which are currently performed by human operators. There is usually a reduction in scrap and die damage when automating this type of application. The simple operation just described is a good example of a basic robotic work cell. Work cells are discussed in detail in Chapter 10.

The T³ Robot is a simple, solidly built 6-axis computer-controlled industrial robot. Unique jointed-arm construction provides the flexibility needed to perform in difficult-to-reach places. Sealed-for-life lubrication and rotary joints with large antifriction bearings mean minimal wear and virtually maintenance-free operation.

Each of the six axes of the T³ Robot is direct-driven by its own electro-hydraulic servo system. The robot is capable of the high torque, speed, and flexibility needed to handle hefty payloads anywhere within an area of up to 1000 cubic feet.

Each axis has its own position feedback device, consisting of a resolver and tachometer, to assure accurate arm repeatability.

The Milacron-built ACRAMATIC computer control provides infinitely variable 6-axis positioning and controlled path (straight line) motion between programmed points. You tell the robot where you want it to go, and it automatically figures out the shortest way to get there. At the velocity *you* select.

To teach the robot its job, you simply use a lightweight, hand-held teach pendant to lead the robot arm through its required moves. Stop points and functions are entered into the control memory with the push of a button. A cathode ray tube is included on the computer control for display of pertinent data.

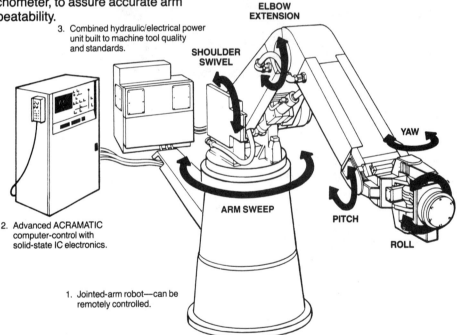

3. Combined hydraulic/electrical power unit built to machine tool quality and standards.

ELBOW EXTENSION

SHOULDER SWIVEL

YAW

ARM SWEEP

PITCH

ROLL

2. Advanced ACRAMATIC computer-control with solid-state IC electronics.

1. Jointed-arm robot—can be remotely controlled.

FIGURE 1.9. Cincinnati Milacron's T³ 566 and T³ 586 sophisticated robot. Courtesy of Cincinnati Milacron Industrial Robot Division, Lebanon, Ohio.

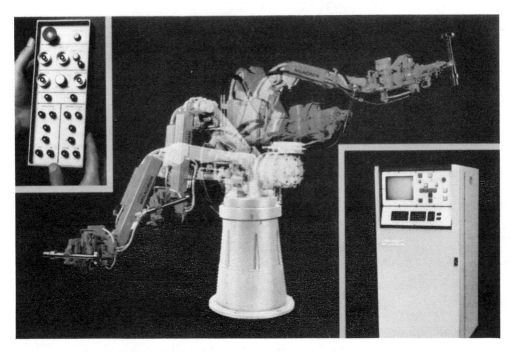

FIGURE 1.9. [Continued].

1.4.2   Die Casting

Figure 1.11 shows a die-casting job similar to, but more complex than, the press-loading application. Die casting is one of the most popular types of medium-technology robot applications.

A Unimate robot is shown servicing a large aluminum die-casting machine, which is producing end housings for electric motors. In this job the robot does the work of one and a half human workers per shift, removing them from the hazard of working between the dies of the casting machine and the trim press. At the same time, the output of finished housings is increased over 10%.

This robot makes use of the special double-handed gripper illustrated earlier in Figure 1.4e. The robot grasps two inserts from the feed chute with one part of the gripper and then moves to the die-casting machine. After unloading the previously completed shot from the machine, the gripper is swung 180° to load the next inserts, still holding the last shot. This avoids much unnessary movement of the robot arm. The completed casting is then placed in a quench tank, and a previous part is removed from the tank to the trim press.

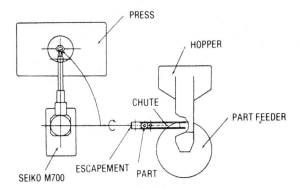

FIGURE 1.10. Seiko M 700 robot loading a press. Courtesy of Seiko Instruments, U.S.A., Inc., Torrence, Calif.

The press is open only about 8 seconds out of the cycle to unload the shot and load the two inserts. The production rate is about 160 shots per hour.[4]

1.4.3   Arc Welding

Welding and welding-related work are one of the areas in which the more sophisticated industrial robots have been making significant inroads. In continuous-path arc welding, for example, robots usually maintain an actual arc time of about 80%, compared to about 25% for human workers.

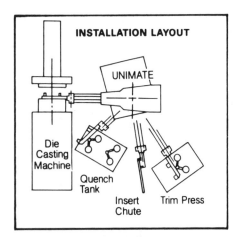

Installation layout showing the Unimate servicing the die-cast machine.

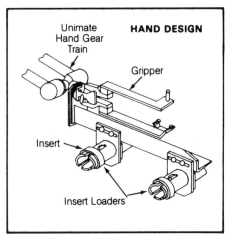

Special hand design employed to load the inserts.

FIGURE 1.11. Die-casting job more complex than press-loading application. Installation layout and hand design are shown. Courtesy of Unimation® Inc., Danbury, Conn.

Figure 1.12 shows a complex work cell for continuous welding. A two-position indexer in front of each robot carries two identical welding fixtures at opposite sides of the turntable. While the workpieces in one fixture are being arc welded by the robot, a human operator between the robots unloads the welded unit from the second fixture and reloads the fixture with the pieces for the next one.

The operator then steps to the adjoining welding station and repeats the unloading and loading operation at the indexer for the second robot. As soon as a robot is finished welding a piece, it activates the indexer, bringing the next unwelded unit into position and also unclamping the just-completed piece from the fixture.

In this application, Unimation reports that labor hours were cut to a quarter of that previously required, contributing to an 18% return on investment. Also, the welds made by the robot were regarded as having a uniformly better appearance than before automation.[5]

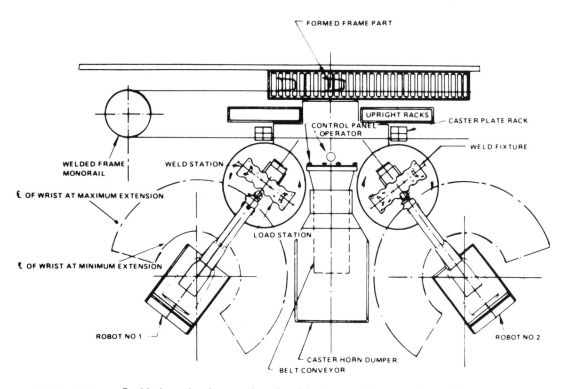

FIGURE 1.12. Sophisticated robot work cell, with two continuous-path welding robots integrated into a complex work cell. Courtesy of Robotic Age, Inc., Petersborough, N.H.

Another characteristic of the most advanced robots is their ability to make use of sensory information to adapt their behavior in response to changes in the work environment.

1.4.4   Deburring and Other Robot Uses

In a system used by ASEA, Inc., an IRB-6 robot, shown in Figure 1.13, can operate a deburring grinder to follow the edge of a casting by means of a sense of touch. Unlike robots that must have the path they are to follow precisely specified in advance in their control programs, the ASEA IRB-6 robot automatically adapts to variations in the position or shape of the workpiece. Its sensor allows it to grind the flange down to match the actual contours of the finished product, much like a human worker who can see the results of his action. The ASEA IRB-60 can also use a touch sensor in conjunction with adaptive control to grind away welding beads by following the bead and "feeling" the actual contour of the workpiece.

The ASEA IRB-6 and IRB-60 robots are also used in the following applications: die casting, forging, investment casting, machine tool loading and unloading, parts transfer, small parts assembly, finishing, plastic molding, arc and spot welding, machining, electronic assembly, inspection gluing, snag grounding, and polishing.

1.4.5   Robot Technology Benefits

Some of the benefits resulting from the appropriate application of robot technology follow:[6]

1. *Increased productivity.* Productivity gains ranging from 10% to 67% have been reported. No slowdown in production on second and third shifts. Uptime as high as 98%, usual with robots, means better utilization of capital equipment. The increased speed of robots also permits realizing the full potential of the high-speed machines they service.
2. *Improved product quality.* Once properly taught to perform a task, a robot will continue to do so repeatedly and consistently, meaning fewer rejects and less scrap. Users have reported improvement in quality control ranging up to 70%.
3. *Flexibility.* Unlike "hard" automation, which requires tremendous overhead in tooling for product changes, robots can be retaught and reassigned to many varied tasks. A robot's gripper and tooling can be changed quickly and easily, and almost any reasonably intelligent person can teach a robot its new task in a matter of hours.
4. *Precision.* Robots move with speed and ease, yet can also place their hands and tools with great accuracy. Repeatable positioning within 0.008 in. is common, and the IBM Cartesian assembly robot has a repeatability of 0.001 in.

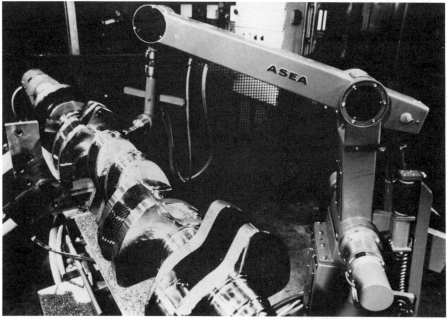

FIGURE 1.13. The ASEA IRB–6 industrial robot. Deburring application is also seen. Courtesy of ASEA, Inc., White Plains, N.Y.

5. Improved morale. It is frequently reported that workers who supervise or work with robots take a new pride in their work and in the progressiveness of their employer. Also, those who have been relieved of dangerous, tiring, boring, dirty, hot, noisy, or otherwise unpleasant jobs often experience a resurgence of pride, dignity, and a sense of loyalty to their company.

To all this are added lower indirect and hidden costs resulting from reduced inventories, longer life from tools and dies, lower average absenteeism in the human work force, and lower costs for OSHA compliance.

1.5   Advanced Robot Systems

Robot work cells of the sort described previously are the basic building blocks of the automated factory. Work cells are linked together to form a "line," along which the processing of workpieces progresses from one step to the next. The result is a flexible, integrated robotic manufacturing system. The link points between the work cells usually incorporate a buffer storage unit of some type to provide some degree of isolation between the cells, permitting shutdown of a work cell for scheduled or unscheduled maintenance or setup of a new process without shutting down the entire line. Work cells upstream would continue to process parts and feed the storage unit serving the inoperative cell. The next cell downstream would continue to draw parts from its own storage unit. This filling and depleting of the storage units would be monitored by a supervisory control computer, which can use each work cell's throughput to optimize the output of the entire line.

Such a robot system can be used to manufacture a family or "group" of parts. This concept is shown as *group technology,* and basically categorizes the parts into families that are of similar size and configuration and require similar processing. As the computer becomes more widely used in manufacturing for scheduling, inventory, and production control, the work flow of a system will become more efficient, and it will be easier to identify the families of parts. Even without using robot systems, dramatic improvements in work flow can be obtained by using group technology. Recent studies have shown that most parts spend up to 90% of the time on pallets, in bins, or being trucked around a plant. Compare that with the continuous work flow made possible by the automated robot processing line, in which the parts are transferred directly from one manufacturing step to the next.

By applying group technology through computer-controlled machinery and robots, manufacturers begin to approach the next higher step in automated production called *computer-aided manufacturing,* or CAM. The U.S.

Air Force is currently sponsoring research and development leading to "integrated CAM" (ICAM). The basic idea of the study is the development of flexible computer-controlled manufacturing cells linked to other cells, forming automated manufacturing lines with a broad range of capabilities. It is believed that CAM and ICAM will significantly alter batch-type manufacturing by the early and mid-1980s.[7]

The hardware for these systems will be made up of computer-controlled machine tools and robots linked to other cells. The cell controllers will in turn be linked to each other and to system controllers, as shown in Figure 1.14. This arrangement of computers into such a system is known as *hierarchical control*. The fact that data are gathered and processed for control and monitoring purposes throughout this network has also led to use of the term *distributed intelligence*. The use of these techniques will make possible the goal of *fully* integrated control of manufacturing facilities.[8]

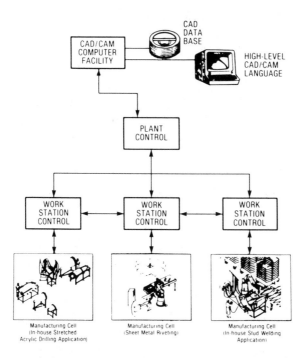

FIGURE 1.14. Advanced robot work cells, each under the control of their own computer, may in turn interact with other cells, under the control of a central plant computer. The plant controller may in turn communicate with CAD facilities. Courtesy of Robotic Age, Inc., Petersborough, N.H.

In related studies, Westinghouse Electric Corporation and the National Science Foundation are participating in a $1.8 million program to research and develop a computer-controlled robotic assembly system for the batch manufacturing of electric motors. When giving reasons for the study, Richard Abraham, Westinghouse manager of programmable automation, said, "The average annual increase in productivity from 1960 to 1977 is only 2.8% in the United States, while it is 8.4% in Japan and 5.7% in West Germany." NSF representative Richard Green said that "unless robotic projects are successful, the nation risks losing labor-intensive manufacturing to countries where wages are lower."[9]

Table 1.1 is a Delphi-method forecast, conducted in 1976 by the Japanese Industrial Robot Association, predicting the technological development of industrial robots and their application in a variety of areas. The results of the first questionnaire predicted the periods of development, initial application, and widespread use in each of the areas. A second group of respondents, given the first results, produced a second set of predictions in the same areas.

In a recent paper stating labor's view of robots and other forms of automation, Thomas L. Weekley, International Representative, The United Automobile, Aerospace, and Agricultural Implement Workers of America, said:

> The UAW membership as a whole does not currently place a specific emphasis on robots. Rather, it views robots as just one of the technological advancements facing the union today. In general, the membership has favored the introduction of such advances and has recognized them as essential in promoting economic progress through increased productivity. Further, labor as a whole has accepted and encouraged these changes.
>
> Not only does increased productivity mean additional benefits to management but employees recognize that their wages, working hours, fringe benefits, and safety will also improve as productivity increases. Productivity, therefore, is a recognized and accepted necessity for advancement of personal and corporate goals. The demand for quality low-cost goods makes increased productivity essential for economic progress.[10]

1.6   Future Developments

By the mid-1980s, several countries will have computer-controlled, automated manufacturing plants using industrial robots. Human workers will perform maintenance, technical, and supervisory functions. If these plants prove to be successful, they may indeed be the beginning of a new industrial revolution—an era of automation, where people are unbound from the machines to which they have been tied since the first industrial revolution.

TABLE 1.1
A DELPHI METHOD FORECAST OF A 1976 STUDY
Source: Robotics Age, Summer 1980, pp. 13–14. Used with permission.

TECHNOLOGICAL SUBJECTS		PERIOD (1976 – 1980 – 1985 – 1990)
1. Modular or built-in Block robots.	Development	▨ ~1980
	Practice application	▨ 1980–1985
	Popularization	▨ 1982–1990
2. Robot Languages for programming.	Development	▨ ~1980
	Practice application	▨ 1983–1985
	Popularization	▨ 1985–1990
3. Machining robots (machine tool robots).	Development	▨ ~1980
	Practice application	▨ 1983–1985
	Popularization	▨ 1985–1988
4. Painting robots for interim painting process of car body.	Development	▨ ~1979
	Practice application	▨ ~1980
	Popularization	▨ 1981–1983
5. Painting robots in final painting process of car body.	Development	▨ ~1980
	Practice application	▨ 1981–1983
	Popularization	▨ 1982–1985
6. Assembling robots which will be applied to assembling processes for small batch production of smaller products.	Development	▨ 1980–1982
	Practice application	▨ 1980–1986
	Popularization	▨ 1984–1988
7. Assembling robots, which will be applied to assembling processes for medium quantity batch production of smaller products.	Development	▨ ~1980
	Practice application	▨ 1982–1984
	Popularization	▨ 1985–1988
8. Assembling robots which will be applied to assembling processes for small batch production of medium size products such as motorcycles.	Development	▨ 1980–1983
	Practice applicaltion	▨ 1983–1986
	Popularization	▨ 1985–1990

TABLE 1.1 (Continued)

TECHNOLOGICAL SUBJECTS		PERIOD (1976 – 1990)
9. Assembling robots for cylinder blocks and carburetors of automobiles.	Development	■ (≈1979–1980)
	Practice application	■ (≈1981–1983)
	Popularization	■ (≈1983–1985)
10. Assembling robots for transmission gear cases of automobiles.	Development	■ (≈1979–1980)
	Practice application	■ (≈1981–1984)
	Popularization	■ (≈1984–1987)
11. Coal mining robots.	Development	■ (≈1980–1981)
	Practice application	■ (≈1984–1986)
	Popularization	■ (≈1985–1988)
12. Maintenance robots for reactor furnaces.	Development	
	Practice application	■ (≈1984–1985)
	Popularization	■ (≈1985–1989)
13. Crop-dusting robots for house cultivations and orchards.	Development	■ (≈1979)
	Practice application	■ (≈1981–1985)
	Popularization	■ (≈1984–1986)
14. Cleaning robots for floors of public buildings and outside walls, windows of high story buildings.	Development	■ (≈1980–1981)
	Practice application	■ (≈1981–1983)
	Popularization	■ (≈1985–1987)
15. Underwater movable robots for geological survey in less than 300m depth.	Development	■ (≈1980)
	Practice application	■ (≈1983–1985)
	Popularization	■ (≈1986–1988)
16. Underwater movable robots for welding machining etc. in less than 300m depth.	Development	■ (≈1980–1981)
	Practice application	■ (≈1982–1985)
	Popularization	■ (≈1985–1992)

For these plants to be successful, improvements will be required in computer hardware and software, robotics, inspection devices, sensory devices, and system design and integration. All these elements are in existence today and rapid advancement in each technology is being made.

The robot is primarily a manipulator-controlled device with software. Without software the robot would be just another machine. At present, in most applications the software controls the arm on repetitive tasks that require precise fixturing or placement of the parts worked on. As the tasks become more challenging (i.e., require the flexibility the human mind is capable of), the software concepts will have to become more sophisticated. Basically, a human being performs a task by drawing on past experience and learning to meet new and changing circumstances. Clearly, to perform more complex tasks, robot software of the future will have to access to some database of its surroundings and make judgments based on that information.

Researchers in the area of artificial intelligence (AI) have already done significant work in the area of database access systems. AI researchers have also done work in decision making and strategy, as witnessed by their game playing programs which can beat all but the best human players. These AI techniques, highly refined and supported by specialized high-speed hardware, will expand the potential applications of robots.

Sensory devices provide information to the robot or work-cell controller. Sensory devices are mounted internally within the robot to provide feedback information on arm position and tactile, force, torque, heat, and other sensorial data. External to the robot, a nearly unlimited amount of devices can be interfaced with the robot to provide required information on the workpiece, environment, process, and ancillary equipment. Inspection devices function as quality controllers. The sensory and inspection devices form their own distributed intelligence network similar to the large system of hierarchical controls.

Vision is our most valued sense. It will play an important role in freeing the robot from its dependence on support equipment. It will add a whole new dimension to robotics.

By using solid-state TV cameras and computers, scientists are now providing robots with "eyes," greatly extending the industrial robot's ability to function in the real world of manufacturing.

Figure 1.15 is a block diagram showing the many applications where vision will assist robots and was prepared by SRI's Artificial Intelligence Center.[11] Figure 1.16 shows a block diagram and data path drawing of a visual-servo control system provided by SRI International's Artificial Intelligence Center.[12] Figure 1.17 shows a more complex system integrating two industrial robots (one simple, one sophisticated) with an xy table, part presenter, TV camera, and other sensory devices.[13] Vision modules like the one shown in Figure 1.16 are now available to industry and are being used for production.

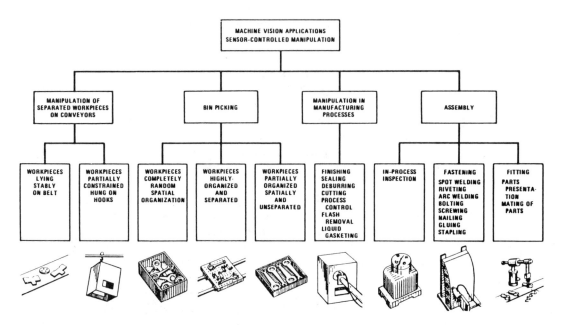

FIGURE 1.15. Block diagram of potential vision applications in robotics. Courtesy of the SRI International's Artificial Intelligence Center, Menlo Park, Calif.

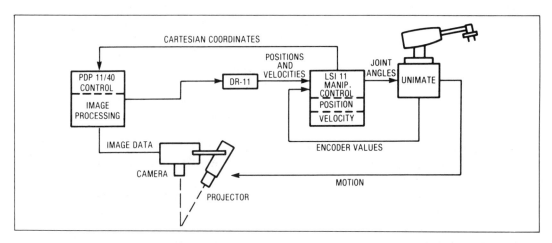

FIGURE 1.16. Block diagram of visual-servo control system. Courtesy of the SRI International's Artificial Intelligence Center, Menlo Park, Calif.

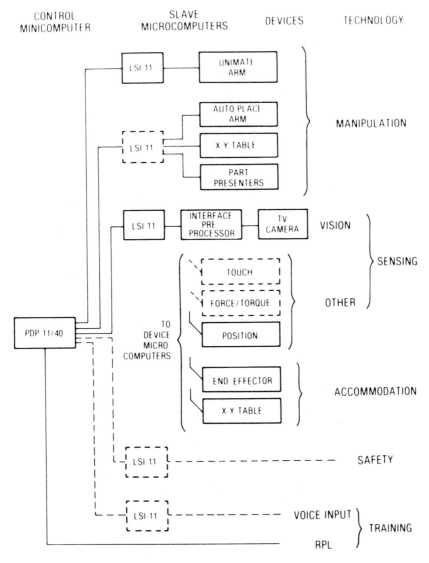

CONTROL
MINICOMPUTER

SLAVE
MICROCOMPUTERS

DEVICES

TECHNOLOGY

LSI 11

UNIMATE ARM

AUTO PLACE ARM

X Y TABLE

PART PRESENTERS

MANIPULATION

LSI 11

INTERFACE PRE PROCESSOR

TV CAMERA

VISION

TO DEVICE MICRO COMPUTERS

TOUCH

FORCE/TORQUE

POSITION

END EFFECTOR

X Y TABLE

SENSING

OTHER

ACCOMMODATION

PDP 11/40

LSI 11

SAFETY

LSI 11

VOICE INPUT

RPL

TRAINING

FIGURE 1.17. Block diagram of distributed computer system for robotic sensing and feedback. Courtesy of the SRI International's Artificial Intelligence Center, Menlo Park, Calif.

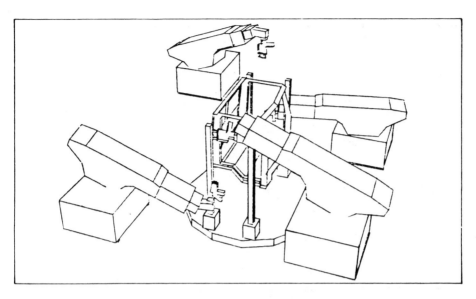

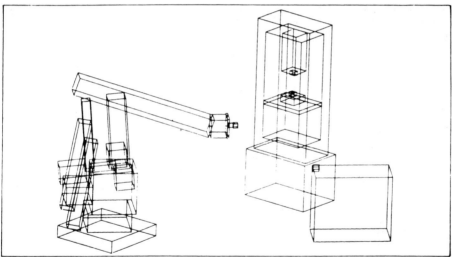

FIGURE 1.18. GRASP, Graphical Robot Applications Simulation Package, software, a computer aided design system planned to ease the introduction of present-day robot technology into the manufacturing environment. The completed system will enable engineers to produce three-dimensional models of robots and workplaces and display them on the screen of a graphics terminal linked to a computer. It will thus be possible to evaluate robot installations and debug robot programs prior to commissioning. It is further proposed to use the knowledge gained from a simulation to define directly the operation of the robot. Two examples of such systems are shown. The top pictorial shows four robots in a body-cab welding process. The bottom pictorial shows a rotary forging process. Courtesy of Department of Production Engineering and Production Management, University of Nottingham, University Park, Nottingham NG7 2RD.

Computer-aided design (CAD), along with computer-aided manufacturing (CAM), are growing in use.

Engineers, architects, and scientists are now able to use the computer as a design tool to sketch objects in three dimensions. In the future it will be possible to connect together computer-aided design equipment with computer-controlled machine tools. This will make it possible for an engineer to design a part on a computer terminal in his office. When satisfied with the design, the engineer will be able to push a button and cause computer-generated control signals to be transmitted to an automatic machine tool, perhaps many hundreds of miles away, where the part will actually be produced without further human intervention.

Computer-aided design is also being used to build mathematical models of robot work cells to help engineers analyze system layout and production rates. This saves enormous amounts of engineering and design time. It also reduces the need to do expensive mock-ups of the task.

Figure 1.18 is an interactive computer graphic drawing of a robot.

All the developments covered in this chapter are already used by industry—or are being tested in laboratories. Because of the increasing rate of change in computer technologies, the most adventurous experts are hesitant to forecast future developments any further than five or ten years. However, studies have been made by the Society of Manufacturing Engineers in this country and by the Japan Industrial Robot Association. Table 1.1 shows the graphic results of a 1976 JIRA study. Most studies of this kind tend to be conservative, and technological growth is projected as being linear. The manufacturing plants will have robots. Robots are here to stay. In the not too distant future, robots will be an integral and workable part of domestic homes.

1.7   Review Questions

1. Give a definition of a robot.
2. Describe the components of a robot now available.
3. Describe the types of industrial robots now available.
4. Discuss five examples of current applications of robots.
5. Discuss advanced robot systems.
6. Discuss how computer-aided manufacturing is related to robots.
7. Draw a block diagram of a visual-servo control system.
8. Draw a diagram of the potential vision applications in robotics.
9. Draw a block diagram of a distributed system for robotic and feedback.
10. Discuss the future of robots and robotics.

1.8 Reference Notes

1. J. S. Albus, *Peoples' Capitalism: The Economics of the Robot Revolution*, U.S. Department of Commerce, College Park, Md., 1976, p. 1.
2. G. Bylinsky, "The Smart Young Robots on the Production Line," *Fortune*, Dec. 17, 1979, p. 90.
3. Kanji Yonemoto, *The Present Status and the Future Outlook of Industrial Robot Utilization in Japan*, Japan Industrial Robot Association, Nov. 1978, p. 1.
4. "Aluminum End Housings for Electric Motor Die Cast by Unimate Robots," *Unimate Industrial Robot Die Casting Casebook*, Danbury, Conn., p. 6; latest publication date 4/81.
5. "Unimate Robots Provide Versatility in Automating Arc Welding," *Unimate Industrial Robot Welding Casebook*, Danbury, Conn., p. 8; latest publication date 4/81.
6. J. Quinian, "Robots Are Taking over Many Jobs Nobody Wants—And Boosting Productivity," *Material Handling Engineering*, Oct. 1979, p. 64.
7. McDonnell Douglas Corporation, "Production Integration Plan, Robotic System for Aerospace Batch Manufacturing," IR–812–8, May 12, 1979.
8. S. Kahne, I. Lefkowitz, and C. Rose, "Automatic Control by Distributed Intelligence," *Scientific American*, June 1979, p. 78.
9. T. Stundza, "Automated Batch Assembly Aim of Study," *American Metal Market/Metalworking News*, Feb. 12, 1979.
10. T. Weekley, "The UAW Speaks out on Industrial Robots," *Robotics Today*, Winter 1979–1980, p. 25.
11. C. Rosen, "Machine Vision and Robotics: Industrial Requirements," Technical Note 174, SRI's Artificial Intelligence Center, Menlo Park, Calif. Nov. 1978, p. 36.
12. C. Rose and others, *Machine Intelligence Research Applied to Industrial Automation*, Eighth Report, Aug. 1978, p. 61.
13. D. Nitzan, *Robotic Sensors in Programmable Automation*, Technical Note 183, SRI's Artificial Intelligence Center March 1979, p. 36.

1.9 Bibliography

Joseph F. Engelberger, *Robotics in Practice*, second printing, American Management Association, New York, 1980.

N. Graham, *Artificial Intelligence*, Blue Ridge Summit, Pa., 1979, p. 11.

Mikell P. Groover, *Industrial Robots: A Primer on the Present Technology*, American Institute of Industrial Engineers, Inc., Nov. 1980, pp. 54–61.

W. Heginbotham, M. Dooner, and R. Case, "Rapid Assessment of Industrial Robots Performance by Interactive Computer Graphics," *Proceedings 9th International Symposium on Industrial Robots,* Society of Manufacturing Engineers, Dearborn, Mich., Mar. 1979, p. 573.

Rich Merrit, "Industrial Robots: Getting Smarter All the Time," *Instrument and Control Systems,* July 1982, pp. 32–37.

John J. Obraut, "Robotics Extends a Helping Hand," *Iron Age,* Mar. 19, 1981, pp. 53–83.

Terence Thompson, "Robots for Assembly," *Assembly Engineering,* July 1981, pp. 32–37.

2

Minicomputers, Microcomputers, and Microprocessors

2.1 Introduction

The evolution of computers has taken the state of the art of electronics from large general-purpose data-processing machines that filled several rooms to a single-chip computer that contains the equivalent of over 100,000 transistors. This revolution in computer construction is the key element that will make robots more and more versatile as time passes. The low cost of single-chip computers means that a robot could have many processors, each devoted to a special task. All contemporary industrial robots have microprocessor-based features.

The essential function of any computer is to take a set of data and perform a sequence of operations upon those data. Traditionally, there are two classifications of this essential task. The first is referred to as data processing (i.e., the data are entered through a terminal, punched cards, microfilm, etc.), and then the data are manipulated by the computer program. An example of this would be analysis of census data or generating payroll checks. For data processing, the data generally can be manipulated subsequent to data collection or entry. The second classification is generally called real-time processing. In real-time analysis, data are collected and analyzed almost instantaneously, and usually an appropriate action is generated by the computer in response to the data. An example of this would be attitude control of a satellite by gathering of data from onboard sensors and then generating appropriate commands to the attitude jets. While the robot computer is basically a real-time controller, it could also do data processing (e.g., by keeping track of parts quality in a machine-loading operation for later generation of statistical reports).

Until the development of the microprocessor minicomputers were invariably used for real-time control. In response to the invasion by microprocessors, minicomputer manufacturers have developed cost-effective single-board versions of their popular minicomputers. This has blurred somewhat the distinction between minicomputers and microcomputers. This distinction is blurred even further by the introduction of single chips, which can emulate the minicomputers. As far as the robot designer or user is concerned, this seesaw battle between mini- and microcomputer manufacturers is and will continue to be a positive situation, since it results in a more powerful and cost-effective robot.

2.2   The Components of a Computer

Figure 2.1 shows the component blocks of a computer. The computer contains memory, a central processing unit (CPU), and mass storage or peripherals.

The memory section contains the program that the CPU must execute, as well as the data needed for program execution. Today, 99% of memory resides on tiny solid-state chips. The last few years have seen an astonishing increase in memory density. In 1975, chips with 256 memory locations were the norm; today chips with 256,000 memory locations are available. Chips with millions of memory locations will exist in the near future. Such memory capacity will be needed as robots become more sophisticated.

The block labeled CPU provides timing, temporary storage areas, and the instruction set of the computer. The power of the CPU is determined by the speed of instruction execution, the variety or richness of the instruction set, the manner in which instructions are handled, and the size of the memory it can access. The trend in microprocessors is toward CPUs with improved technologies for higher execution speeds, greater and greater variety in the instruction set, and new architectures that allow for processing of several in-

FIGURE 2.1. Components of a computer.

structions at once. For example, the Intel 8086, a 16-bit microprocessor, has 135 instructions, an instruction cycle time of roughly 1 microsecond (μs) and the ability to address 1 million memory locations. All this power is expected to cost less than $40 in large quantities.

The input/output section of the computer provides an interface to the world outside the computer, permitting the connection of devices such as keyboards, displays, switches, and relays. Many I/O chips are quite complex and versatile; in fact, some are specialized computers. This sophistication can remove part of the burden from the CPU, enabling the CPU to perform many different tasks within the robot.

The mass-storage section of the computer provides an area to store large blocks of data, as well as a place for very large programs. Common mass-storage devices are magnetic tapes, disks, video tape, and bubble memories. Mass storage provides a large area for keeping data and programs at a significantly lower cost than solid-state main memory. Mass storage has also seen a tremendous increase in storage densities, with a dramatic attendant drop in cost per storage location.

Future robots will need these large mass-storage capabilities for applications such as maneuvering in a space where the robot must remember a map of the area in which it is working.

2.3　Brief Description of the Subsystems of a Computer[a]

The memory is a unit for the storage of instructions and data, and it contains a large number of cells or locations. Each position is used to store n bits of binary information. The information stored in a location constitutes a *word* of n *bits* (binary digits); a word of 8 bits is called a *byte*.

The precision of a calculation is dependent on the number of bits used to represent a number and therefore on the size of the words. The size of the words varies from one machine to another; for a minicomputer it ranges from 8 to 24 bits. The size of the memory is equal to the number of locations it contains. In general, it is a multiple of 1K words (1K word = 1024 words of n bits) because a modular design is nearly always adopted.

Each location of the memory is identified by a unique *address*, which allows access to the contents of the location. Consequently, to obtain an item of

[a]Sections 2.3 through 2.5 are extracted in part from Harve Fireford, *Introduction to Microprocessors* 1st ed., Motorola Semiconductor Products, Inc., © 1975. Courtesy of Motorola Semiconductor Products, Inc., Phoenix, Arizona.

information from the memory, it is necessary to know the address at which this information is stored. At any given time it is only possible to address a single location. A memory of 8K words may be represented as shown in Figure 2.2.

The types of memory include the following:

1. Semiconductor memories
2. Magnetic memories
   a. Core memory
   b. Drum memory
   c. Disk memory
   d. Magnetic tapes

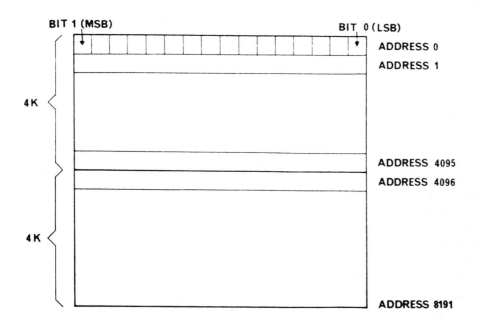

MSB is the most significant bit. LSP is the least significant bit.

FIGURE 2.2. Representation of a memory of 8K words. Courtesy of Motorola Semiconductor Products, Inc.

2.3.1   Interfacing between the Memory and the CPU

Communication between memory and the CPU is achieved by means of two buses and two registers and includes the address bus and address register for the addresses and the data bus and data register for the data, that is, for the *contents* of the memory locations specified by the address.

When the CPU wants to read the information in a given location, it puts the address of the location in one of its internal registers, the address register, and then transfers the contents of this register AR to the memory on the address bus. After decoding the address, the location selected sends its contents to the CPU on the data bus. These contents are stored in another register of the CPU termed the data register.

The same principle applies to writing; the contents of the data register are stored at the address indicated by the address register. The mode of access to a memory position is termed *random* because it allows immediate access to any memory position; in contrast, there also exist memories with *sequential* access (magnetic disks). A memory that permits random access is termed a *random-access memory* (RAM) or *read/write memory*.

Certain memories only allow reading; the information is frozen in and cannot be destroyed by overwriting. These are *fixed memories* or *read-only memories* (ROM).

The memory of a minicomputer can be composed of both ROMs and RAMs together. In the ROMs may be stored the permanent utility programs, such as the programs for reading a punched tape from a high-speed or slow reader. Constant values may also be stored here, while the RAMs will contain the variables of the problem.

Example.   Calculation of the pay of an empoyee: the rates for an hour of normal work and for an hour of overtime can be stored in the ROM, while the variable values which are the number of hours of each type done by the employce, will be stored in the RAM.

The processor or CPU shown in Figure 2.3 is an active part of a computer. The CPU is composed of several subsystems, of which the most important are the following:

1. *Accumulator register ACC* in which takes place the arithmetic and logic operations.
2. *Link or carry register or flip-flop.* This is a register of 1 bit, which is considered as an extension of the accumulator; it is used in particular to connect via a loop the MSB and the LSB of a number contained in the accumulator during rotation operations.
3. *Status register,* which is composed of five special flip-flops:
   a. The carry flip-flop just mentioned, which is affected either by rotation operations, by arithmetic, or by logical operations giving rise to a carry.

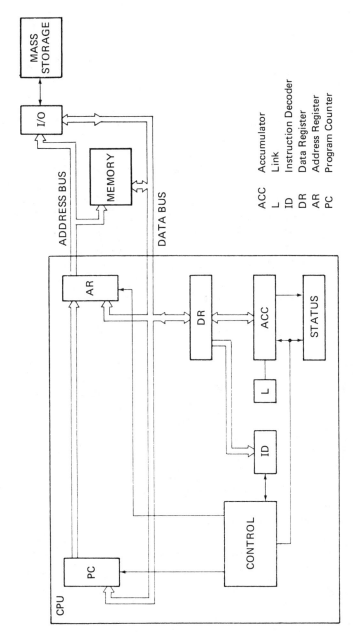

FIGURE 2.3. Block diagram of the subsystems of a computer. Courtesy of Motorola Semiconductor Products, Inc.

ACC   Accumulator
L   Link
ID   Instruction Decoder
DR   Data Register
AR   Address Register
PC   Program Counter

b. The negative flip-flop, which is set to 1 when the contents of the accumulator become negative.

c. The zero flip-flop, which is set to 1 when the contents of the accumulator become zero.

d. The overflow flip-flop, which is set to 1 when an operation causes overflow on the sign bit of a number.

e. The interrupt flip-flop, which permits a request for interruption to be granted or not.

4. *Instruction decoder register* (ID) in which is stored the operation code part of an instruction; an instruction is composed of many bits, such as 8, 16, or 24 bits. For example, the 4 most significant bits can indicate the operation code such as that for addition and the other bits the address for the operand.

5. *Program counter register* (PC) contained in the CPU. The computer memory contains the program to be followed, that is, a series of instructions. Consequently, at any time it must know which instruction is to be carried out, or rather at which address the instruction is stored in the memory. The role of the PC is to hold this address. At the end of each cycle, the PC indicates the address of the next instruction to be carried out; this thus allows the sequential execution of a program of which the instructions are also sequentially stored in the memory. However, in some cases the contents of the PC may be modified by the program itself. In this way it is possible to carry out instructions stored in another part of the program. This is done by branching instructions.

6. *The control unit* is also an integral part of the CPU. This unit coordinates all the parts of the computer in such a way that the events take place in the correct sequence and at the right time.

The I/O system provides the interface between the processor and the outside (peripheral unit). To do this, it must be able to solve:

1. The problem of timing; peripheral equipment is generally much slower than a computer.

2. The problem of the format of the information transmitted to the computer (series–parallel translation).

3. The problem of hardware; a peripheral unit does not necessarily have the same logic as the computer.

Having reviewed the different subsystems of the computer, we are now ready to describe its structure.

The three main parameters that characterize a computer system for robots are the size of the memory, the speed of execution of the instructions, and the number and type of instructions available.

2.3.2 Input/Output Transfer Functions

Microcomputers find their applications in industrial robotics, numerical control, data acquisition, and digital electronic measurements. For these applications, it is essential that the computer communicate with the peripherals of the control and measurement system.

There are at least three ways of exchanging information with the peripherals. The first method is called polling. The second method is achieved through the use of interrupts. The third method discussed in this section is the direct-memory access.

The I/O transfer operation controlled by the microcomputer is called *programmed I/O operations* or *polling*. The following criteria are required for the proper operation of polling. To carry out a programmed I/O transfer, it is necessary to provide communication with a peripheral to both receive and test the information describing the state of the peripherals, as well as have the ability to send an item of data from the computer to the peripherals and to receive an item of data from the peripherals. These conditions dictate the structure of a peripheral, or rather the CPU–peripheral interface. This interface must be provided with a status register allowing control of the peripheral and a data register allowing exchange of the data.

Conditional transfer or polling is widely used and takes place under the control of the program when the peripheral under the control of the program is ready to communicate. This method consists of testing (or polling) at regular intervals the status register of the peripheral and waiting until the data are available. When the status register indicates that this is so, the computer transfers data to and reads it from the data register of the peripherals. The main disadvantage to conditional transfer lies in the fact that it is necessary to periodically interrogate the peripheral to see if it is available, resulting in loss of time to the computer.

The most efficient type of transfer saving computer time is the transfer using interrupts. This method is also controlled by a program, but the computer does not have to continuously test the status register of the peripheral and wait for it to become available. In the normal way, the computer can execute a program called the "background job." When the peripheral is ready to effect a transfer, it asks to interrupt the computer, making its request on a special line of the processor termed the *interrupt request line*. The CPU then interrupts its background job and indicates to the other peripherals that it is about to perform an interrupt. The CPU indicates this status by setting to 1 the interrupt bit of the status of the CPU. Then the CPU, which has momentarily dropped its background job, will carry out the transfer routine appropriate to the peripherals. The CPU is said to "service" the peripheral. Once the transfer has taken place (using the programmed service routine), the CPU resets the interrupt bit to 0 and continues with its background job at the point where it left off. In many applications, several peripherals are connected to the same computer. In a typical configuration, all interrupt request lines are OR-wired to the single interrupt request line of the computer. The computer must identify the peripheral that has requested the interrupt by assigning priorities to the interrupt requests of the peripherals. When the peripheral requests an interrupt, it sets to 1 a special flip-flop system associated with it called a *flag flip-flop*. Each peripheral has its own flag.

Direct-memory access (DMA) is a type of I/O transfer operation not controlled by a computer program. The transfer of a block of data takes place without the control of the processor. The peripheral can therefore obtain direct access to the memory via the address and data buses. DMA is found in systems equipped with peripherals working with blocks of data requiring high-speed transfer (e.g., disk storage systems).

2.4   The Binary System

Most quantities in nature are of a continuous nature; for example, weight, position, and speed are measured on scales that ideally have an infinite number of points. In contrast, digital systems are not continuous; rather, they have discrete values. This often requires special circuits to convert the discrete signals of the computer to the robot's requirements of analog position and speed.

Computers are digital in nature because they are based on the binary number system. The binary number system has only two digits, 0 and 1, as opposed to the decimal system, which has 10 digits, 0 to 9. The rationale for using the binary system in computers is that it is *very* easy to fabricate tiny, high-speed electronic switches that have only two states, on or off. Each of these tiny switches can be used to represent one binary bit. Therefore, a computer with thousands of binary bits has thousands of electronic switches.

In a digital system, each signal line represents one binary bit. For example, Figure 2.4 shows the output of a binary counter. The digital lines are labeled A, B, C, and D.

Since there are four output lines, this chip is called a 4-bit binary counter. The reader should note that the binary system is based on the powers of 2 system and is very easy to work with. As an example,

$$[1010]_2 = 2^3 + 2^1 = 10$$
$$[0111]_2 = 2^2 + 2^1 + 2^0 = 7$$

The trigger shown in Figure 2.4 is a series of digital pulses. Each pulse causes the counter to advance one space.

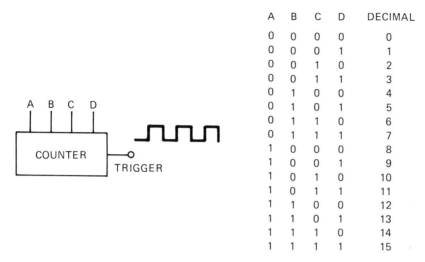

A	B	C	D	DECIMAL
0	0	0	0	0
0	0	0	1	1
0	0	1	0	2
0	0	1	1	3
0	1	0	0	4
0	1	0	1	5
0	1	1	0	6
0	1	1	1	7
1	0	0	0	8
1	0	0	1	9
1	0	1	0	10
1	0	1	1	11
1	1	0	0	12
1	1	0	1	13
1	1	1	0	14
1	1	1	1	15

FIGURE 2.4. Four-bit counter and truth table.

Every computer must be able to represent numbers and carry out operations on the numbers thus represented. From childhood, we have been accustomed to representing quantities in the decimal system, that is, in a system said to be "of *base* 10"; the digits of this system run from 0 to 9. By giving these digits different positions or *weights*, it is possible to represent numbers greater than 10. In the general case of a system of base *b*, the successive positions of the digits, from left to right, have the following weights:

$$\ldots b^3 b^2 b^1 b^0 \cdot b^{-1} b^{-2} \ldots$$

The symbol "." (radix point) separates the integral part from the fractional part of the number. A computer uses the system of base 2 (binary system) rather than base 10 simply because most physical systems have only two alternative states (in particular, the digital components of the computer).

The decimal value of a binary number is obtained by multiplying each bit by its weight and adding the products. Consider the conversion of the binary number of 10011.10.

$$10011.10 = 1.2^4 + 0.2^3 + 0.2^2 + 1.2^1 + 1.2^0 + 1.2^{-1} + 0.2^{-2} = 19.5$$

Decimal–binary conversion can be best explained by an example. We will convert the decimal number 53.5625 into binary notation (Figure 2.5). The integral part is converted by successive division by 2 and the fractional part by successive multiplication by 2.

The memory of the computer is organized in bytes (words of 8 bits). With 8 bits, it is possible to represent a quantity by a binary number between 0 and

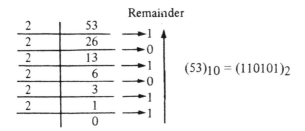

Finally: $(53.5625)_{10} = (110101.\,10010)_2$

FIGURE 2.5. Decimal–binary conversion. Courtesy of Motorola Semiconductor Products, Inc.

11 111 111 = 377_8 or, in the decimal system, between 0 and 255_{10}. Base 8 is called the octal system. By convention, the MSB (bit 7) is taken to represent the sign of the quantity whose value is indicated by the bits 0 to 6.

In general, the following statements are true:

1. MSB = 0 for positive numbers.
2. MSB = 1 for negative numbers.
3. The highest positive number will be 01 111 111 = 177_8 = $+127$.
4. The highest negative number (in absolute value) will be 10 000 000 = $200_8 = -128$.

To represent a negative number, it is only necessary to observe the following rules:

1. Represent the absolute value of the negative number in pure binary notation.
2. Invert all the bits (complement with respect to 1).
3. Add 1 to the LSB.

Any binary subtraction $A - B$ may be reduced to the addition of the number $(-B)$ to A. To obtain the number $(-B)$, it suffices to take the two's complement of B.

The binary coded decimal code (BCD) is a hybrid notation in which each digit is replaced by its pure binary equivalent in 8421 code (on 4 bits). As an example, consider 93_{10}.

Example: $93_{10} = [1001\ 0011]_{BCD}$. Conversely, $[0100\ 0110]_{BCD}$
$= 4 \times 10^1 + 6 \times 10^0 = 46_{10}$.

On 8 bits, it is possible to represent $2^8 = 256$ numbers. The maximum precision that it is possible to obtain in a calculation is therefore $1/256$. When it is decided to represent a number by a single word, it is said to be represented in *single precision*. If the precision obtained is not sufficient, representation in multiple precision may be used, with the number stored in several words. There is also another way of representing numbers: *floating-point* notation, in which the mantissa and the exponent of the number are distinguished.

A further benefit of the digital system used for robots is that many components used in robotics are two-state devices (e.g., switchs, relays, photodetectors, and ac motors). Each digital line can either control or monitor one of these components. For example, a Unimation PUMA robot has 48 digital I/O lines, and hence can control or monitor 48 switches, relays, motors, and the like.

Just as in hydraulic systems, where valves are used to control the flow of fluid to different parts of the system, elements called *gates* are used to control the signal flow in digital systems. Figure 2.6 depicts the two basic gates: the AND gate and the OR gate.

The truth tables in Figure 2.6 defining the gates show that the AND gate requires both input signals to be a 1 to obtain an output (hence the name—AND) while the OR gate only needs one input to be a 1.

To illustrate the use of the gates as control elements, Figure 2.7 shows an AND gate used to control the flow of trigger signals to a counter. If the line

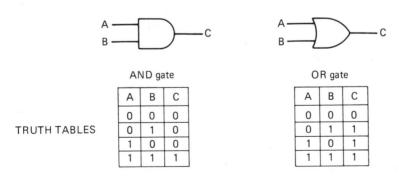

AND gate

OR gate

A	B	C
0	0	0
0	1	0
1	0	0
1	1	1

A	B	C
0	0	0
0	1	1
1	0	1
1	1	1

TRUTH TABLES

FIGURE 2.6. AND gate and OR gate and truth tables.

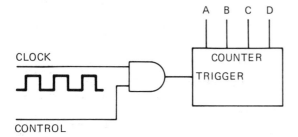

FIGURE 2.7. AND gate used to control a counter.

labeled control is high (a binary 1), the clock signals will pass through the gate and cause the counter to advance. On the other hand, if the control signal is low (a binary 0), the clock signal will not pass through and the counter will not advance.

2.5   Microprocessors

Microprocessors (MPU or μP) are often referred to as "MOS/LSI computers on a chip." MOS/LSI is the abbreviation for metal-oxide semiconductor for large-scale integrated circuits. This rather dangerous label calls for a word of warning: the microprocessor and the associated components needed for its operation must certainly not be regarded as the "miracle product" capable of equaling or surpassing the performance of a computer. In most cases the microprocessor can advantageously replace a wired logic system of some importance. The μP of a robotic system can be interfaced to the external connections via a data bus or other interface system.

The principle of the MP is to perform the functions programmed in its read-only memory; this fixed program, which cannot be modified, is termed *firmware*. This principle differs, on the one hand, from that of general-purpose computers programmed by software and, on the other hand, from that of digital systems, which can only perform the function that was wired into them. Different programs may be stored in the read-only memory (ROM) of the MP system to cover the following applications:

1. Point of scale terminal
2. Inventory control
3. Telecommunications control
4. Chemical analysis
5. Control of machine tools
6. Multiprocessor or microcomputer
7. Bank terminal

 8. Intelligent terminal
 9. Medical electronics
 10. Optical character recognition

All these may someday involve robots for a particular application. The potential for industrial robotics in these areas is great and many uses are already a reality.

2.6   The Intel 8085AH/8085AH–2/ 8085AH–1 8-Bit HMOS Microprocessors[b]

The Intel 8085AH (Figure 2.8) is a complete 8-bit parallel central processing unit (CPU) implemented in N-channel, depletion mode, silicon gate technology (HMOS). Its instruction set is 100% software compatible with the older 8080A microprocessor, and it is designed to improve the 8080A's perfor-

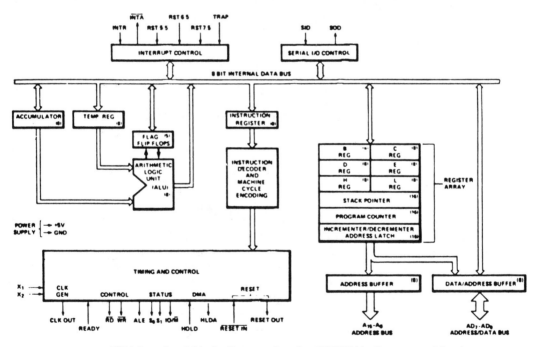

FIGURE 2.8. CPU functional block diagram for the 8085AH. Courtesy of Intel Corporation.

[b]The material in Section 2.6 is from Intel Application Notes APN–01242C, April 1981, and is used courtesy of the Intel Corp.

mance by higher system speed. Its high level of system integration allows a minimum system of three ICs [8085AH (CPU), 8156H (RAM/IO), and 8355/ 8755A (ROM/PROM/IO)], while maintaining total system expandability. The 8085AH–2 and 8085AH–1 are faster versions of the 8085AH.

The 8085AH incorporates all the features of the clock generator and the system controller chips provided for the older 8080A, thereby offering a high level of system integration. The 8085AH uses a multiplexed data bus. The address is split between the 8-bit address bus and the 8-bit data bus. The on-chip address latches of 8155H/8156H/8355A memory products allow direct interface with the 8085AH.

The features of these microprocessors include the following:

- Single +5-V power supply with 10% voltage margins.
- 3-, 5-, and 6-MHz selections available.
- 1.3-μs instruction cycle (8085AH); 0.8 μs (8085AH–2); 0.67 μs (8085AH–1).
- 100% software compatible with 8080A.
- On-chip clock generator (with external crystal, *LC* or *RC* network).
- On-chip system controller; advanced cycle status information available for large system control.
- Serial in–serial out port.
- Decimal, binary, and double precision arithmetic.
- Direct addressing capability to 64K bytes of memory. Such microprocessors are useful in industrial robotic systems.

2.6.1   Programming

There is much more concentration on the awareness of input and output from the microprocessor. The programmer must be adept at dealing with the processor ports and setting them up for entry and exit of data. The choice of a suitable microprocessor may well be strongly influenced by the simplicity with which this can be done. For example, the Intel 8080 and related types, although offering somewhat less versatility than others, have a considerably simpler port setting procedure than, for example, the older Motorola 6800. Programming will be explained in more detail in Chapters 8 and 9.

The time factor must be most carefully observed. Input quantities that vary rapidly must be examined more frequently than those that do not or vital information may be lost. In a closed-loop control system, too infrequent sampling of input data can also lead to instability.

All computer programs are really just lists, but when inputs have to be constantly monitored, they become circular lists repeating over and over.

Sometimes microprocessor inputs change in a completely unpredictable way or it may be that an alarm input must be recognized immediately. In this case, most microprocessors use hard-wired interrupts to "take over" the computer temporarily. Programmers need to know how to handle these events and how to use the facilities provided for assigning relative priorities to one or more such inputs.

It should be remembered that each robotic manufacturer provides its own robot programming languages. To date there are more than 14 such language systems. Programmable microprocessor-based controllers can use ladder diagrams instead of languages to facilitate industrial electricians.

2.7   A Single-Chip Microcomputer System

While microprocessors are generally used for high-level computer applications, single-chip microcomputers are used for low-level applications. Such single-chip microcomputers are used in robotic systems and other control applications. These microcomputers contain memory and I/O devices on a single chip, thus providing space savings and local control functions that can be incorporated in one overall system.

Some 8-bit microcomputers commonly used include the Motorola MC6801 and 6805, Intel 8031/8051, and Fairchild 3870. In Sections 2.8 and 2.9, the Motorola MC6801 and Intel 8031/8051 are described.

2.8   Motorola MC6801
Microcomputer/Microprocessor[c]

The MC6801 is an 8-bit single-chip microcomputer unit (MCU) that significantly enhances the capabilities of the original M6800 family of parts. It includes an upgraded M6800 microprocessor unit (MPU) whose instructions are compatible with the older 6800 chips. Execution times of key instructions have been improved, and several new instructions have been added, including an unsigned multiply. The MCU can function as a self-contained microcomputer or can be connected to as much as 64K of external memory. It is compatible with other logic chips and requires one +5-V power supply. On-chip resources include 2048 bytes of ROM, 128 bytes of RAM, a serial communications interface (SCI), parallel I/O, and a three-function programmable timer. An electrically programmable read-only memory (EPROM) version of the MC6801, the MC68701 microcomputer, is available for systems development.

---

[c] The material in Section 2.8 is from Motorola Application Notes DS9841, © Motorola, Inc., 1981, and is used courtesy of Motorola Semiconductors Products, Inc.

Figure 2.9 gives the pin assignments for the MC6801. Figure 2.10 provides the block diagram of the M6801 microcomputer family.

The MC6801 MCU family features include the following:

- Enhanced MC6800 instruction set
- 8×8 multiply instruction
- Serial communications Interface (SCI)
- Software compatibility with the M6800
- 16-Bit three-function programmable timer
- Single-chip or expanded operation to 64K byte address space

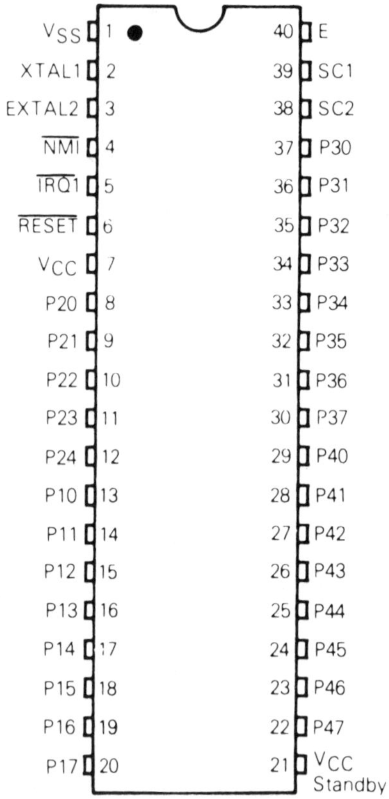

```
     VSS  [ 1  ●        40 ]  E
   XTAL1  [ 2            39 ]  SC1
  EXTAL2  [ 3            38 ]  SC2
     NMI  [ 4            37 ]  P30
    IRQ1  [ 5            36 ]  P31
   RESET  [ 6            35 ]  P32
     VCC  [ 7            34 ]  P33
     P20  [ 8            33 ]  P34
     P21  [ 9            32 ]  P35
     P22  [ 10           31 ]  P36
     P23  [ 11           30 ]  P37
     P24  [ 12           29 ]  P40
     P10  [ 13           28 ]  P41
     P11  [ 14           27 ]  P42
     P12  [ 15           26 ]  P43
     P13  [ 16           25 ]  P44
     P14  [ 17           24 ]  P45
     P15  [ 18           23 ]  P46
     P16  [ 19           22 ]  P47
     P17  [ 20           21 ]  VCC
                              Standby
```

FIGURE 2.9. Pin assignment for the MC6801. Courtesy of Motorola Semiconductor Products, Inc.

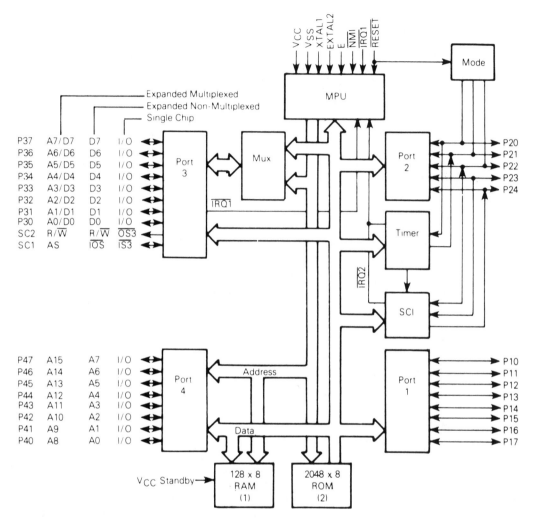

FIGURE 2.10. M6801 microcomputer family block diagram. Courtesy of Motorola Semiconductor Products, Inc.

- Bus compatibility with the M6800 family
- 2048 Bytes of ROM (MC6801)
- 128 Bytes of RAM (MC6801 and MC6803)
- 64 Bytes of RAM retainable during power down (MC6801 and MC6803)
- 29 Parallel I/O and two handshake control lines
- Internal clock generator with divide-by-4 output

2.9   Intel 8031/8051 Single-Component 8-Bit Microcomputer[d]

The Intel 8031/8051 is a stand-alone, high-performance, single-chip computer fabricated with Intel's highly reliable + 5-V, depletion-mode, N-channel, silicon-gate HMOS technology. It provides the hardware features to make it a powerful and cost-effective controller for applications requiring up to 64K bytes of program memory and/or up to 64K bytes of data storage.

The 8051 contains read-only program memory; read/write data memory; 32 I/O lines; two 16-bit timer/counters; a five-source, two-priority-level, nested interrupt structure; a serial I/O port for either multiprocessor communications, I/O expansion, or full duplex UART; and on-chip oscillator and clock circuits. The 8031 is identical with the 8051 except that it lacks the program memory.

The 8051 microcomputer is efficient both as a controller and as an arithmetic processor. The 8051 has extensive facilities for binary and BCD arithmetic and excels in bit-handling capabilities. Efficient use of program memory results from an instruction set consisting of 44% one-byte, 41% two-byte, and 15% three-byte instructions. With a 12-MHz crystal, 58% of the instructions execute in 1.0 μs, 40% in 2.0 μs, and multiply and divide require only 4.0 μs. Among the many instructions added to the standard 8048 instruction set are multiply, divide, subtract, and compare.

The block diagram, logic symbols and pin configuration for the Intel 8031/8051 microcomputer are shown in Figures 2.11 to 2.13.

The Intel 8031 microcomputer contains a control-oriented CPU with a RAM and an I/O, while the Intel 8051 consists of an 8031 with a factory-masked programmable ROM.

The CPU chip in Figure 2.11, which executes the program, can control a robot. The CPU also controls the information to the robot actuators and the peripheral equipment. The basic communication path for the 8031/8051 microcomputer to other parts of the robotic system is picked up on the data and address buses. The data and address buses are labeled in Figure 2.12 as PORT O.

Digital data flow to and from the CPU chip via eight lines. For example, if the 8031/8051 microcomputer wished to check if any of the robot limit switches were accidentally hit, this information would appear on the data bus.

Microcomputer cost and power are always important in a robotic system.

---

[d]The material in Section 2.9 is from Intel Corporation Preliminary Application Notes 210395–002, August 1982, and is used courtesy of Intel Corporation.

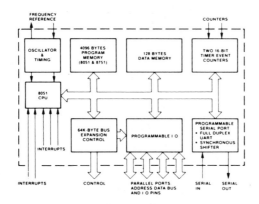

FIGURE 2.11. Block diagram of the Intel 8031/8051 single-component 8-bit micro-computer. Courtesy of Intel Corp.

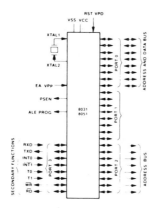

FIGURE 2.12. Logic symbol of the Intel 8031/8051. Courtesy of Intel Corp.

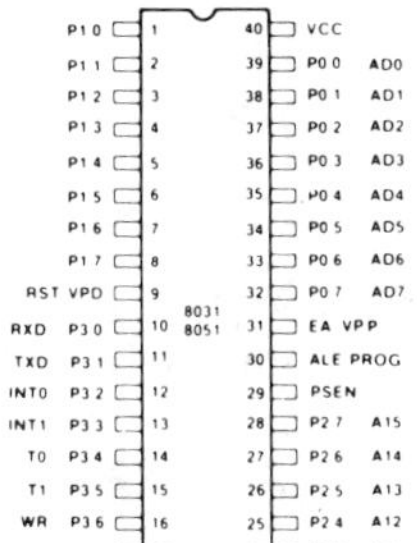

FIGURE 2.13. Pin configuration of the Intel 8031/8051. Courtesy of Intel Corp.

2.10   Intel 80/30 Single-Board Computer[e]

The Intel's powerful 8-bit N-channel 8085A CPU, fabricated on a single LSI chip, is the central processor for the iSBC 80/30. The 8085A CPU is directly software compatible with the Intel 8080A CPU. The 8085A contains six 8-bit general-purpose registers and an accumulator. These registers may be addressed individually or in pairs, providing both single- and double-precision operators. The minimum instruction execution time is 1.45 μs. The 8085A CPU has a 16-bit program counter. An external stack, located within any portion of iSBC 80/30 read/write memory, may be used as a last-in, first-out storage area for the contents of the program counter, flags, accumulator, and all six general-purpose registers. A 16-bit stack pointer controls the addressing of this external stack. This stack provides subroutine nesting bounded only by memory size.

The iSBC 80/30 has an internal bus for all onboard memory and I/O operations and a system bus (i.e., the Multibus) for all external memory and I/O operations. Hence, local (onboard) operations do not tie up the system bus and allow true parallel processing when several bus masters (i.e., DMA devices, other single board computers) are used in a multimaster scheme. A block diagram of the iSBC 80/30 functional components is shown in Figure 2.14.

The Intel 80/30 contains two microprocessors: an 8085 acting as a master CPU and an 8041 single-chip microprocessor acting as a slave or intelligent-I/O processor.

The iSBC 80/30 contains 16K bytes of dynamic read/write memory using Intel 2117 RAMs. All RAM read and write operations are performed at maximum processor speed. Power for the onboard RAM may be provided on an auxiliary power bus, and memory protect logic is included for RAM battery backup requirements. The iSBC 80/30 contains a dual-port controller, which provides dual-port capability for the onboard RAM memory. RAM accesses may occur from either the iSBC 80/30 or from any other bus master interfaced via the Multibus. Since onboard RAM accesses do not require the Multibus, the bus is available for any other concurrent operations (e.g., DMA data transfers) requiring the use of the Multibus. Dynamic RAM refresh is accomplished automatically by the iSBC 80/30 for accesses originating from either the CPU or via the Multibus. Memory space assignment can be selected independently for onboard and Multibus RAM accesses. The onboard RAM, as seen by the 8085A CPU, may be placed anywhere within the 0 to 64K address space. The iSBC 80/30 provides extended addressing jumpers to allow the onboard RAM to reside within a 1-megabyte address space when accessed via the Multibus. In addition, jumper options are provided that allow the user to reserve 8K- and 16K-byte segments of onboard RAM for use by the 8085K

---

[e]The material in Section 2.10 is from Application Notes 9800/100A, 1978, and is used courtesy of the Intel Corp.

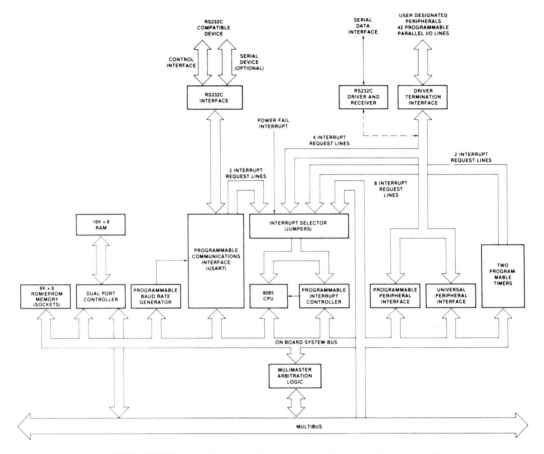

FIGURE 2.14. iSBC 80/30 single-board computer block diagram. Courtesy of Intel Corp.

CPU only. This reserved RAM space is not accessible via the Multibus and does not occupy any system address space.

The three buses in 80/30 hierarchy shown in Figure 2.14 are an onboard bus, a dual-port (DP) bus, and the Multibus (system bus). Innermost is the on-board bus, which connects the 8085A, all onboard I/O peripherals, and ROM. The next bus in the hierarchy, a dual port, connects a dual-port controller, 16K bytes of dynamic RAM, and a dynamic RAM controller. The outermost bus, the Multibus, offers modules that permit either the expansion or addition of system resources. Multibus and iSBC are registered trademarks of the Intel Corporation.

If a 16-bit single-board microcomputer is required, Intel has manufactured the ISBC 86/05™, which has expanded performance.

2.11   Microprocessor Interfacing
with a Robot[f]

Robot manufacturers use microprocessor-based systems that use control systems in their products. Let us consider two examples to illustrate the importance of the microprocessor.

An electric control system with feedback is used to take full advantage of the power and the versatility of a robotic machine. Feedback of positional information is a feature included in these machines, which, although essential for reliable repeatability, is absent in the majority of all other low-cost machines, severely restricting their use as emulators of the mainstream industrial robots. The block diagram of the microprocessor-based control system in the robotic control system is shown in Figure 2.15.

The heart of this system is the microprocessor unit, which performs the task of overseeing the complete system and issuing control signals in response to arm positional data, programmed-in commands, commands from the control box, or information from an external computer. With its CMOS memory it stores the positions of the arm, and by using servo techniques will instruct the robot to repeat the sequence of tasks programmed in. Each arm and joint position is defined by an 8-bit word giving a resolution of 1 part in 256 (0.4%). Data are transferred to and from the interface board by 8 bits in parallel, while data are received from the control box serially, permitting a simple infrared link to be used for controlling the Genesis M101 mobile machine. In the interests of standardization, interfacing with an external computer is via an RS–232–C serial interface with a range of baud rates between 300 and 9600. Although this facility is included, an external computer is by no means necessary as the system's own processor and operating system provide all the functions normally required.

The robot arm is fed to an interface board consisting of an oscillator detector, power supply, latches, analog-to-digital converter (ADC or A/D), and solenoid driver, as shown in Figure 2.11. The interface board essentially serves many purposes. It generates a precision reference sine wave for the positive detector units. It also detects the feedback data and converts them to a dc voltage, which is then multiplexed and converted to a digital signal via an analog-to-digital converter. All data into and out of the microprocessor are held in latches on this interface board.

The microprocessor board shown in Figure 2.16 is separate and is not used for other applications. It consists of a 6802 microprocessor with a clock generator and 128 bytes of RAM.

---

[f]The material in Section 2.11 is used courtesy of Feedback, Inc., Berkeley Heights, N.J.

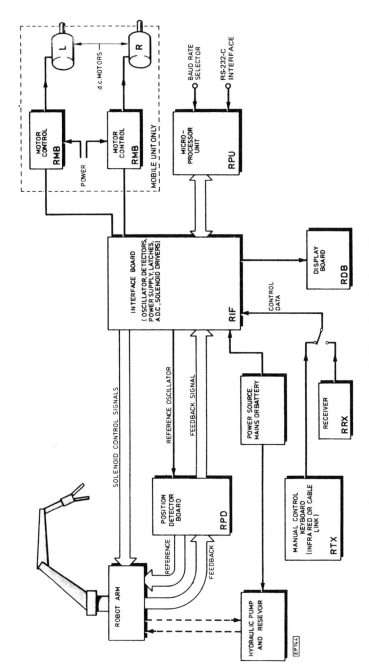

FIGURE 2.15. The robot's general control system in block diagram form. Courtesy of Feedback Inc., Berkeley Heights, N.J.

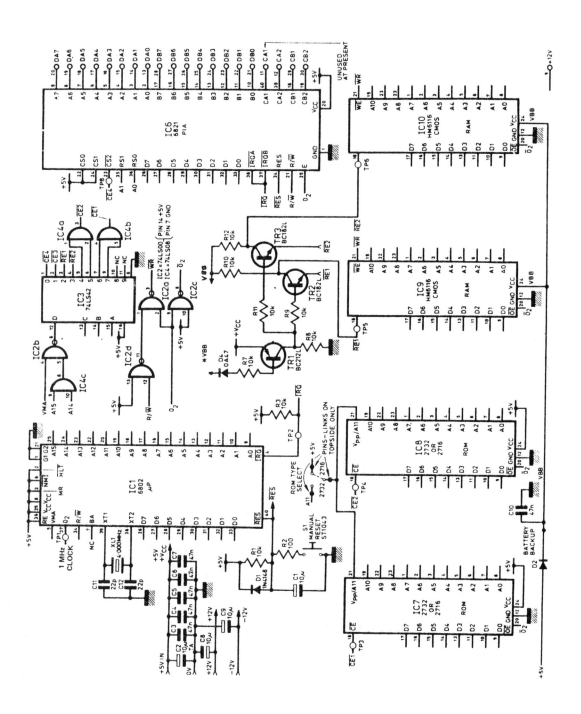

58

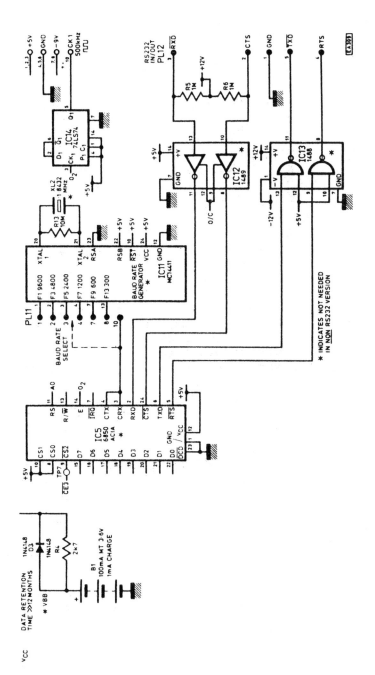

FIGURE 2.16. Processor board circuit diagram. The RAM section is backed up by a rechargeable battery so that software will be nonvolatile. Courtesy of Feedback Inc., Berkeley Heights, N.J.

2.11.1   Control Cabinet of the ASEA IRB 60/2[9]

The control cabinet shown in Figure 2.17 contains the electronic and drive equipment required to control the mechanical robot and the peripheral equipment. The robot is linked to the control cabinet by means of a cable, which may be up to 15 m in length. This is the second example of a microcomputer system.

Communication between the operator and the control system takes place through the control panel and the programming unit. The control system communicates with the peripheral equipment through input/output units, which are available in a number of different variants. A wide selection of number and type of inputs and outputs is available, depending on the needs of the particular installation. The status of the inputs and outputs (I/O) is shown by indicator lamps (LEDs) on the fronts of the I/O units.

The control program for the robot is stored in the permanent memory of the control system (EPROM), while the user program is stored in the read/write memory (RWM), which is provided with battery backup to protect the contents in case of power failure. The capacity of the user memory can be easily

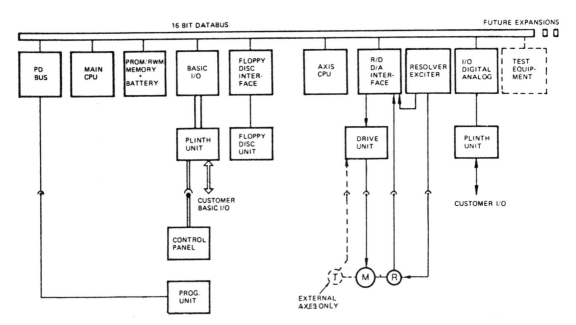

FIGURE 2.17. Control cabinet of the ASEA IRB 60/2. Courtesy of ASEA, Inc., White Plains, N.Y.

[8] The material in Section 2.11.1 is used courtesy of ASEA, Inc., 4 West King Street, White Plains, N.Y.

increased by additional memory boards or by use of a floppy disk unit as a mass storage.

The electronic units in the control system are connected to a common data bus through which all information is passed between units and through which the system is monitored by the central processing unit (CPU).

The servo system is served by its own computer; this contributes to the good performance of the system.

The control system carries out continuous tests on the most important functions of the robot system, and more extensive tests are performed after start-up. If a fault is detected, program execution is immediately interrupted, the fault lamp lights up, and a message showing the type of fault is displayed in plain text on the programming unit.

To assist in fault tracing, the various units of the control system are fitted with test outlets and LEDs that indicate the status of the input and output signals of the system, the servo system, and the interlock chains. There are also LEDs for indication of supply voltages to the electronic unit and for their communication with the central unit.

In addition, separate test equipment can be used in the form of a test adapter and a bus text unit.

2.12   Review Questions

1. Discuss the properties of a microcomputer.
2. Discuss the properties of a microprocessor.
3. Define a RAM, ROM, I/O, and the bit.
4. Discuss the interface systems used in microcomputers.
5. How and why are microprocessors and/or microcomputers used in robotic systems?

2.13   Bibliography

John A. Allocca and Allen Stuart, *Electronic Instrumentation*, Reston Publishing Co., Reston, Va., 1983, Chapter 15.

William S. Bennett and Carl F. Evert, Jr., *What Every Engineer Should Know about Microcomputers*, Marcel Dekker, Inc., New York, 1980.

Ron Bishop, *Basic Microprocessors and the MC 6800*, Hayden Book Co., Inc., Rochelle Park, N.J., 1981.

Herbert Bruner, *Introduction to Microprocessors*, Reston Publishing Co., Reston, Va., 1982.

Walter Buchsbaum and Gina Weissenberg, *Microprocessor and Microcomputer Data Digest*, Reston Publishing Co., Reston, Va., 1983.

Thomas Davis, *Experimentation with Microprocessor Applications,* Reston Publishing Co., Reston, Va., 1981.

Joseph D. Greenfield and William Gray, *Using Microprocessors and Microcomputers: The MC 6800 Family,* John Wiley & Sons, Inc., New York, 1981.

ISBC 80/30™ *Single Board Computer Hardware Reference Manual,* Intel Corp., Santa Clara, Calif., 1978.

ISBC 88/44™ *Measurement and Control Computer Hardware Reference Manual,* Intel Corp., Santa Clara, Calif., 1981.

Microprocessor Courses, Nos. 1 through 12, Medical Electronics, 2994 West Liberty Ave., Pittsburgh, Pa.

Systems Data Catalog, Intel Corp., Santa Clara, Calif., 1982.

3

The Servo Control
System of a Robot

3.1   Introduction

Servo systems are an integral part of a robotic or a numerical control system. In this chapter, feedback theory is discussed. Open-loop and closed-loop systems, stability, graphic methods, compensation, nonlinearity, and the digital servo system are highlighted.

3.2   Introduction to Feedback
Control Systems[a]

Control systems with feedback take on many forms, but the electromechanical servo is an excellent example of the general system. Figure 3.1 shows such a system in block form, together with a simple static analysis to show that, provided the feedback is in the right sense and that the loop gain is sufficiently high, the output can be made to follow the input more or less exactly.

---

[a] The material in Sections 3.2 to 3.9 appears here courtesy of Feedback, Inc., Berkeley Heights, N.J.

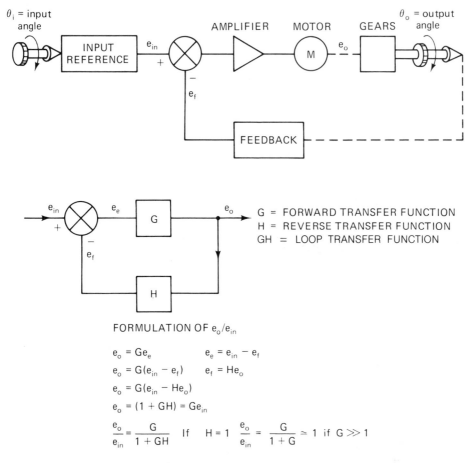

θ_i = input angle

AMPLIFIER   MOTOR   GEARS

θ_o = output angle

INPUT REFERENCE

e_{in}

$+$

$-$

e_f

M

e_o

FEEDBACK

e_{in} e_e G e_o

$+$ $-$

e_f

H

G = FORWARD TRANSFER FUNCTION
H = REVERSE TRANSFER FUNCTION
GH = LOOP TRANSFER FUNCTION

FORMULATION OF e_o/e_{in}

$e_o = Ge_e$ $e_e = e_{in} - e_f$

$e_o = G(e_{in} - e_f)$ $e_f = He_o$

$e_o = G(e_{in} - He_o)$

$e_o = (1 + GH) = Ge_{in}$

$\dfrac{e_o}{e_{in}} = \dfrac{G}{1 + GH}$ If $H = 1$ $\dfrac{e_o}{e_{in}} = \dfrac{G}{1 + G} \simeq 1$ if $G \gg 1$

FIGURE 3.1. A generalized feedback control system.

Most simple analyses of control systems make the assumption that all elements operate in a linear fashion, but in practice many forms of nonlinearity occur. These include deadband, hysteresis, backlash friction, and striction. Figure 3.2 illustrates some of these effects.

Control systems are basically a means of causing a variable to follow a demand, but in the process the system is usually subjected to various types of disturbance, either as a result of spurious changes in an otherwise constant quantity (e.g., power supply) or as a result of loads placed on the system output. The types of disturbance can be numerous and may indeed be quite random, but for the purpose of deciding how well a system can cope with disturbances, a number of simpler types of disturbance are normally considered. Figure 3.3 shows the most important of these. Examples are not difficult to think of; for instance, a box dropping on a conveyor belt represents a step dis-

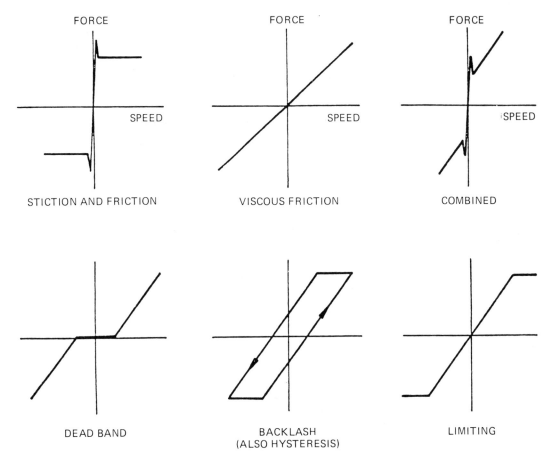

FORCE — SPEED — STICTION AND FRICTION

FORCE — SPEED — VISCOUS FRICTION

FORCE — SPEED — COMBINED

DEAD BAND

BACKLASH (ALSO HYSTERESIS)

LIMITING

FIGURE 3.2. Various nonlinearities.

turbance, while an automobile suspension may be subject to all kinds of disturbance depending on the road surface; but impulses will be quite common, and sinusoidal disturbance may well represent fairly accurately an undulating road surface.

The ability of a control system to reduce its own error to zero depends very greatly on the actual form of the forward transfer characteristic (G in Figure 3.1). Without going into mathematical detail, if G contains an integral term, the error for a given finite output must necessarily be zero, whereas if it does not, there must always be a finite error for a finite output. Systems can contain more than one integral term, in which case different characteristics are obtained, but the commonest are type 0 (no integral term) and type 1 (one integral term). These are illustrated in Figure 3.4.

Analysis and testing of control systems depend to some extent on the final application and on the purpose of the particular test (e.g., is it to establish

t = 0

UNIT IMPULSE

$\text{AREA} = \frac{1}{2} h \quad \frac{2}{h} = 1$

LIMIT AS $h \to \infty$

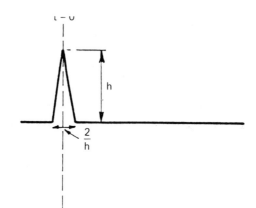

h

$\frac{2}{h}$

UNIT STEP

$= \int (\text{UNIT IMPULSE}) dt$

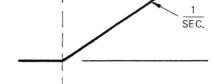

1

UNIT RAMP

$= \int (\text{UNIT STEP}) dt$

$\frac{1}{\text{SEC.}}$

UNIT SINUSOID

f = FREQUENCY (Hz)

SIM $2\pi FT = \text{SIN } \omega t$

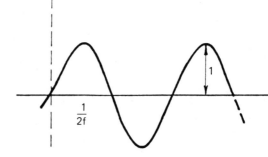

1

$\frac{1}{2f}$

FIGURE 3.3. Examples of input electronic disturbances.

TYPE 0	TYPE 1

NO INTEGRAL IN G

e.g. (i) $G = \dfrac{K}{s + a}$ $H = 1$

$$\frac{e_o}{e_{in}} = \frac{K}{s + a + K} = \frac{K}{a + K}$$

WHEN S = 0 1st ORDER

(ii) $G = \dfrac{K}{(s + a)(s + b)}$ $H = 1$

$$\frac{e_o}{e_{in}} = \frac{K}{s^2 + s(a + b) + ab + K}$$

$$= \frac{K}{ab + K}$$

WHEN S = 0 2nd ORDER

ONE INTEGRAL IN G

e.g. $G = \dfrac{K}{s(s + a)}$ $H = 1$

$$\frac{e_o}{e_{in}} = \frac{K}{s^2 + sa + K} = 1$$

WHEN S = 0 2nd ORDER

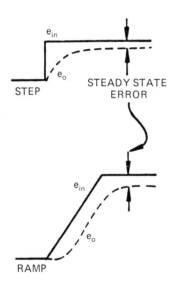

STEP

STEADY STATE ERROR

RAMP

1st ORDER RESPONSES

POSITION ERROR = 0

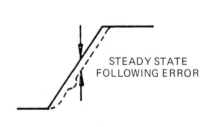

STEADY STATE FOLLOWING ERROR

2nd ORDER RESPONSES

FIGURE 3.4. The meaning of type and order in control systems.

final performance or is it to assist in the design?). The two basic methods used in testing are (1) frequency domain, which uses sinusoidal testing signals, and (2) time domain, which uses various transient-type signals such as step or impulse. One widely used instrument that can be used for both testing methods is the spectrum analyzer.

In the frequency domain, performances are represented by either the Bode or Nyquist diagrams; in the time domain a time graph is the usual representation. Figures 3.5 and 3.6 illustrate these concepts. Either of these two graphical methods may be used to represent both the open-loop and closed-loop response. By open-loop response is meant the behavior of the for-

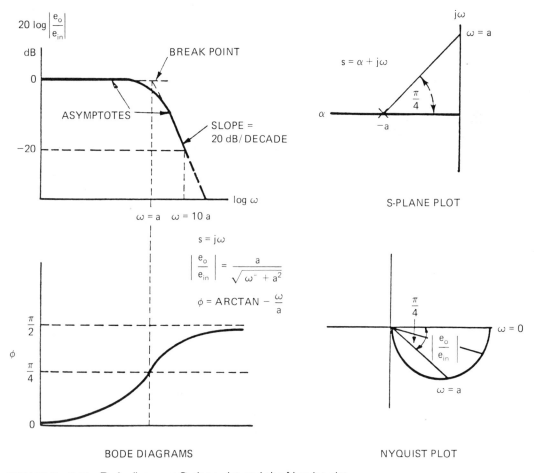

FIGURE 3.5. Bode diagrams S-plane plot and the Nyquist plot.

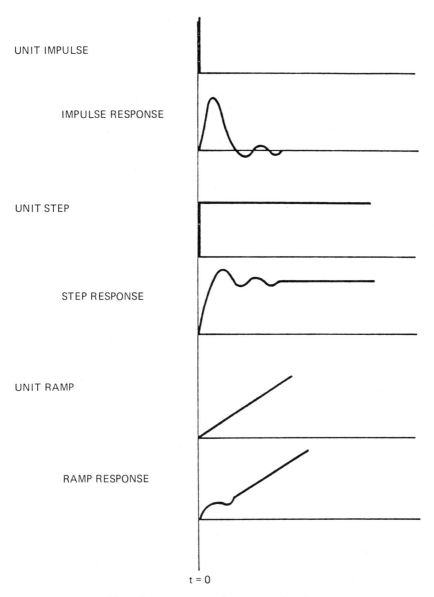

UNIT IMPULSE

IMPULSE RESPONSE

UNIT STEP

STEP RESPONSE

UNIT RAMP

RAMP RESPONSE

t = 0

FIGURE 3.6. Transient responses for a second-order system.

ward and reverse transfer characteristics in the absence of a loop closure; that is, how does the quantity GH respond? The closed-loop response is the behavior of the final system once the loop has been closed and is the characteristic associated with the quantity $G (1 + GH)$.

3.3   Open- and Closed-Loop Systems

Usually the open-loop responses are measured and the corresponding graphs used to decide whether or not the closed-loop system will be (1) stable and (2) of satisfactory response. If one or both of these is not true, then compensation may be required, and this will usually be done using the open-loop plots. The robot arm can be either open- or closed-loop depending on the application.

3.4   Stability

The question of stability is determined by some quite simple criteria (actually fairly involved, but they can be applied in a simple fashion). To understand these, we must understand the mechanism of instability.

In Figure 3.1 the quantity GH is called *loop gain*, and it is the variation of this loop gain with frequency that determines whether the loop will be stable or not. If GH is negative and less than unity, any small disturbance in the system will propagate around the loop, ending up with a smaller disturbance, and so will eventually die out. The system is stable. If GH is negative and greater than unity, the same will apply unless at some frequency the phase shift caused by GH is sufficient to turn a negative feedback into a positive feedback. Then a small disturbance becomes a large one and instability results.

From this simple analysis it is easy to deduce that the point at which GH equals unity and at which its phase shift is 180° is a critical point in either of the graphical representations. Figure 3.7 illustrates this for the Bode and Nyquist diagrams.

A typical servo position control system employing a rotary actuator (discussed in Chapter 4) displays the second-order open-loop characteristic, and Figure 3.8 shows the Nyquist plot for such a servo. It is immediately obvious that no matter how high the zero frequency loop gain constant may become, this system can never be unstable. However, as the gain increases and the plot approaches the critical point, the response becomes increasingly oscillatory, with severe overshoots when tested by a unit step.

In the same example, if G contained an additional time constant, which might be due, for instance, to the use of a field-controlled motor or to a more complex type of actuator (e.g., hydraulic), the transfer characteristic becomes third order, as shown, and too much gain very quickly causes instability.

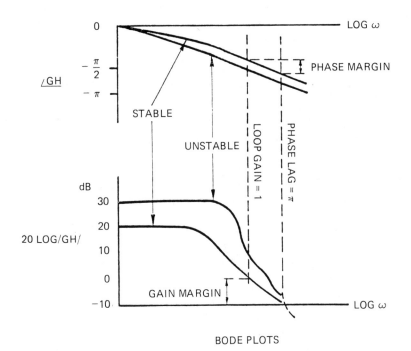

BODE PLOTS

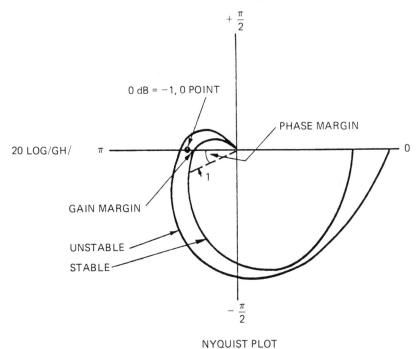

NYQUIST PLOT

FIGURE 3.7. The Bode and Nyquist plot diagram.

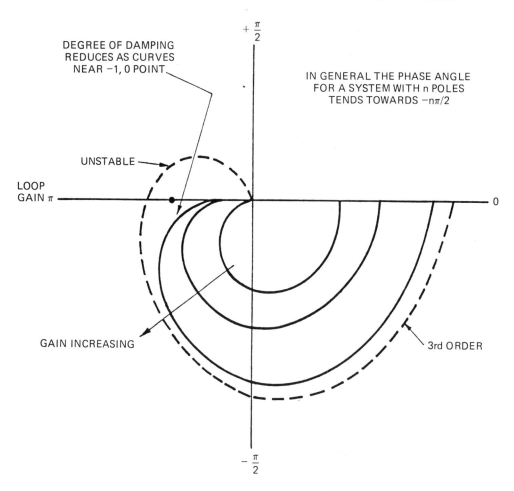

NYQUIST PLOTS OF A TYPICAL
OPEN LOOP TRANSFER — 2nd ORDER

FIGURE 3.8. Nyquist plots of a typical open-loop transfer, second order.

3.5   Suitability of Graphic Methods

The Nyquist plots are most useful in the recognition of incipient instability or underdamping, whereas the Bode plots are most useful for the identification of dynamic parameters when trying to match a theoretical transfer expression to a set of experimental results.

There is a third common method of graphical representation known as the *root locus diagram*. This is not an experimental method, however, and relies upon a knowledge of the mathematical expression for the control system loop characteristics in order to show how the closed-loop behavior will vary as the loop gain is changed between zero and infinity. A typical method of using this diagram is to start with Bode plots of a practical system, to obtain from it an approximation of the theoretical parameters, and to use these in a root locus diagram to assist in the modification of the control loop to obtain the required performance. This brings us to the question of compensation.

3.6   Compensation

Almost any practical system in its rudimentary form will display imperfections of performance of one kind or another, which must be removed or reduced by the application of compensation methods. Some possible faults are as follows:

1. Poor positioning accuracy
2. Poor following accuracy
3. Too slow a response
4. Too many overshoots
5. Too large overshoots
6. Too long settling time

Compensation is achieved in electrical systems by the introduction of frequency-conscious electrical networks that vary the relative phase shift and gain in such a way as to obtain the desired characteristics.

Some common compensation techniques include: limited phase lead, limited phase lag, techogenerator or velocity feedback, and two- or three-term control. The last of these methods is very commonly used in process control and has been implemented for many years in pneumatic form and more recently in electronic form. Figure 3.9 illustrates the application of two- or three-term control to an electromechanical speed control system.

In particular it should be noted that the introduction of an integral term into the forward path converts what is basically a type 0 system into a type 1 system and thus eliminates (at least within the constraints of friction and the like) the steady-state control error.

3.7   Nonlinearity

Most of the preceding discussion has assumed linear operation, but certain nonlinearities are very obvious when they are present and their effects can be intuitively understood without complex theoretical analysis. For example, a

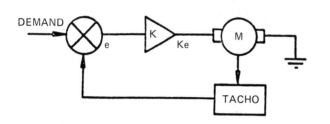

PROPORTIONAL CONTROL

ERROR IS NON-ZERO
HOWEVER HIGH IS GAIN K

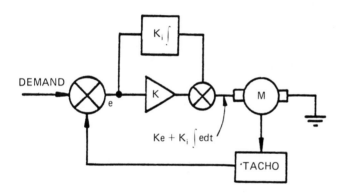

PROPORTIONAL + INTEGRAL

UNLESS e IS ZERO MOTOR
SPEED MUST INCREASE,
WHICH IT DOES UNTIL e
IS ZERO BUT CAUSES
OVERSHOOT ON STEP

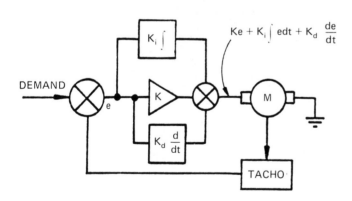

PROPORTIONAL + INTEGRAL
+ DERIVATIVE

DERIVATIVE TERM (DAMPING)
REDUCES OVERSHOOTS

A POSITION SERVO HAS BUILT-IN INTEGRATION SO WHEN TACHO (DERIVATIVE)
f/b IS ADDED BECOMES A 3 TERM CONTROL SYSTEM IN EFFECT.

FIGURE 3.9. Simple speed control.

type 1 position control system, as we have seen, should have no steady-state position error, but in practice if the brush friction of a dc motor is large compared with the normal operating torque of the motor, there may be a residual position error since the motor will stop when its torque falls to the friction level.

Some other forms of nonlinearity are very difficult to assess in their effects upon performance. Backlash in particular may well cause an otherwise stable system to become unstable.

3.8   Microprocessors in Feedback Control Interfacing

Microprocessors and microcomputers (discussed in Chapter 2) have been widely used in feedback systems. The function of microprocessors in feedback control is twofold: monitoring, that is, seeing what is going on, and control, that is, supervising what is going on. Obviously, the first of these involves measurements and, as for the second, if control is to be a closed-loop process (feedback control), then measurement is no less important since it is axiomatic that nothing can be closely controlled unless it can be measured.

Figures 3.10 and 3.11 show the basic principles of open- and closed-loop control systems. In the simplest closed-loop system, we find ourselves measuring and controlling the same variable, but things are not always as simple as this, and Figure 3.12 shows an example of a system that is far from simple. In fact, it is not immediately obvious whether this is an open- or closed-loop system. It is, in simplified form, a fuel-injection system for a modern automobile, and basically one might say that there is a closed-loop manually controlled system setting the engine speed with an open-loop automatically controlled system ensuring that the air-to-fuel ratio is correct for all engine speeds. In the process, numerous measurements are made, and at least one variable, the fuel ratio, is controlled. In addition to these loops, there will be certain overrides for cold starting conditions and other unusual circumstances.

The example in Figure 3.12 emphasizes the need for practically trained technicians to maintain such systems. Their robotic education must include (1) a general background understanding of all the measurement and control techniques, including digital, electronic, and microcomputing, that find a place in such systems and (2) a particular (you might almost say vocational) knowledge of the individual systems on which they may have to work.

Such systems as described are today controlled by microprocessors. They give great flexibility to adapt to different requirements, such as changing laws in industrial robotics.

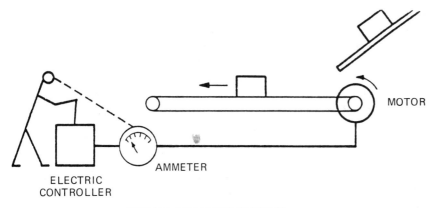

ELECTRIC
CONTROLLER

AMMETER

MOTOR

MANUAL OPEN-LOOP CONTROL

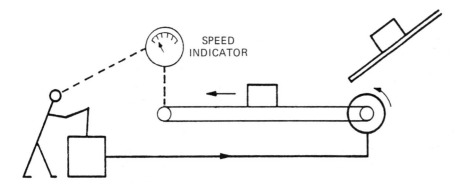

SPEED
INDICATOR

MANUAL CLOSED-LOOP CONTROL

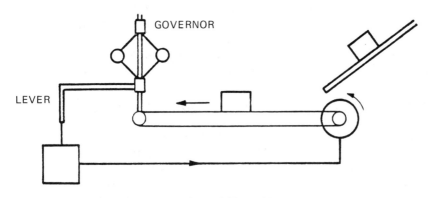

GOVERNOR

LEVER

AUTOMATIC CLOSED-LOOP CONTROL

FIGURE 3.10. Open- and closed-loop control.

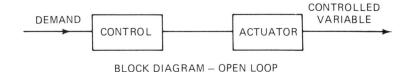

BLOCK DIAGRAM – OPEN LOOP

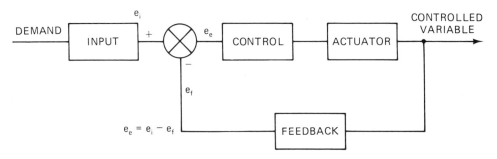

FIGURE 3.11. Block diagram of an open and closed loop with no feedback and with feedback.

Microprocessor and other electronic systems need interfaces wherever they contact the outside world. Here we are not talking about keyboards, video displays, line printers, tape recorders, or even synthesized speech. These are all man–machine interfaces and indisputably important in their right place, but where human intervention is not appropriate or not desired the interfaces must be of a different nature.

Every input (measurement) and every output (actuation) needs its own interface, and the variety of such interfaces is as wide as that of the inputs and outputs themselves. Robot actuators are discussed in Chapter 4.

Most transducers measure physical quantities to provide continuous analog outputs. However, this is not convenient for a digital system. Hence, one aspect of interfacing of vital importance is analog-to-digital conversion. Conversely, most actuating devices for control purposes are also analog, and therefore some form of digital-to-analog conversion is needed. (See Chapter 5.)

The broadest range of techniques is undoubtedly applicable to the treatment of inputs from measured variables; these can arise from a multitude of physical, electrical, and other phenomena applied in such a way as to produce electrical signals that represent the changing variable. In between the transducer itself and the analog-to-digital conversion may be one of the many types of signal-conditioning arrangements. These include bridge circuits, amplifiers, impedance matching, IM links, range matching, and many others.

Figure 3.13 shows some of the types of variables used in a data-acquisition system. In general an input requires a transducer, a signal conditioner, a transmission medium, and an analog-to-digital converter. These elements are covered in Chapter 5.

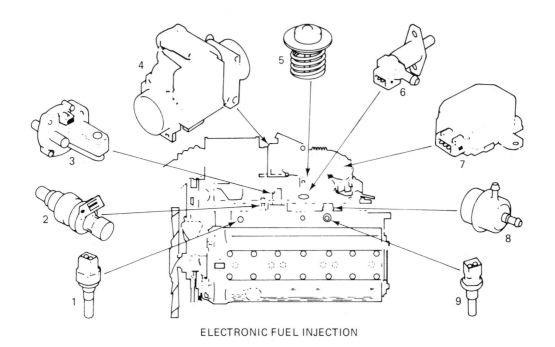

ELECTRONIC FUEL INJECTION

VISUAL FEEDBACK VIA DRIVER

DESIRED SPEED — THROTTLE OPENING (AIR CONTROL) — ENGINE — ENGINE SPEED — TRANSMISSION — ROAD SPEED

AUX AIR (3)   COLD START INJR (6)   INJECTORS (2) ← FUEL

AIR FLOW (4)
COOLANT TEMP (9)
THERMO-TIME (1)
THROTTLE OPENING (7)
ENGINE CRANKING

ELECTRONIC CONTROL UNIT

FIGURE 3.12. An automobile fuel-injection control.

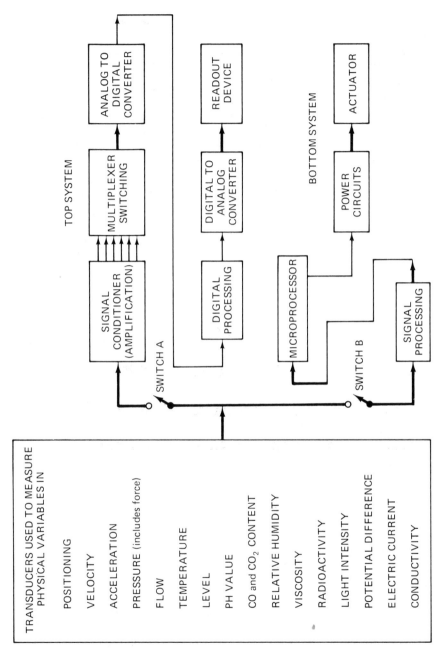

FIGURE 3.13. Block diagram of a data-acquisition system.

NOTE With Switch A in the circuit, the top system can be used. With Switch B in the circuit, the bottom system can be used.

79

Most outputs require more power than is generally available from low-power microprocessors and other digital circuits, so the usual complement will be digital-to-analog conversion, power amplification, and the actuator. Figure 3.14 illustrates the general types of actuators that are used.

Complex control systems like the fuel-injection system of Figure 3.12 require measurement of many quantities, and often the economic viability of the whole system will be set not by the microprocessor or the electronics, but by the cost of transducers and actuators, which are notoriously expensive devices. The most obvious methods for measurement and actuation are thus not necessarily the ones adopted in practice, and there is a constant constraint in the engineering of these systems for simplicity and economy in these areas.

As a simple example, the actuator in the fuel-injection system is basically a continuous controller adjusting the rate of fuel flow, but a pure analog device for that purpose would be expensive and possibly unreliable. Instead, the strategy is adopted of using a simple solenoid on/off control of duty ratio (that is ratio of on to off time), which is controlled by a pulse rate modulated on/off signal from the controller. This gives DAC and power amplification in one instrument.

3.9   A Practical Closed-Loop Control System and Testing

Some time has to be spent discussing the fundamental properties of simple closed-loop systems because without this we cannot understand the constraints within which the microprocessor system must operate, nor can we recognize faulty operation when we see it. Figures 3.15 and 3.16 show two fur-

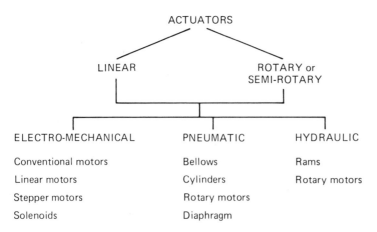

FIGURE 3.14. The actuators of a robotic system.

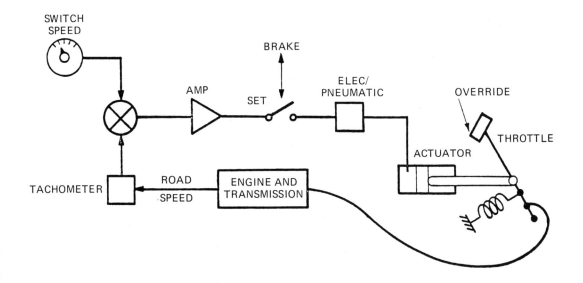

CLOSED LOOP CRUISE CONTROL (PRINCIPLE)

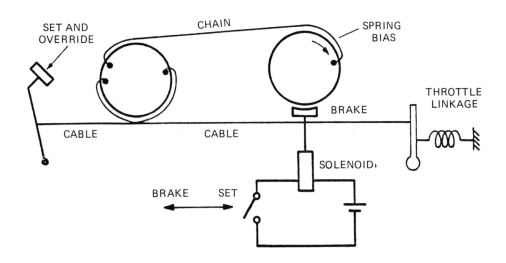

OPEN LOOP CRUISE CONTROL (PRINCIPLE)

FIGURE 3.15. Two automobile speed controls.

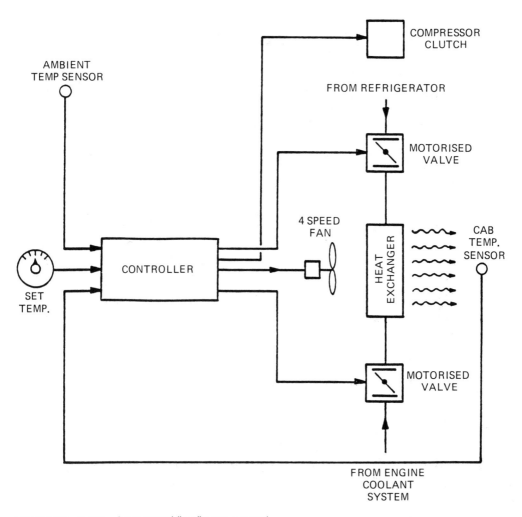

FIGURE 3.16. An automobile climate control.

ther examples from the automobile field that are closed-loop systems, one controlling road speed and the other air temperature and humidity. The relative complexity of the climate-control system yields three inputs and four outputs.

Figure 3.17 shows a much simpler one-input, one-output position control. The diagram shows the controller as hardwired logic, but this can just as easily be replaced by a microprocessor, which can do the same job and also perform other functions, such as generating an arbitrary programmable series of position demands. Notice in this system the elements we have already met, that is, two inputs, one basically numerical and one analog with an analog-to-

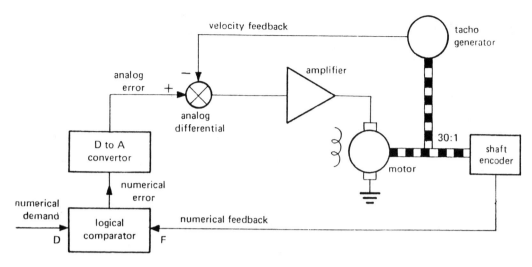

FIGURE 3.17. Numerical position control scheme.

digital converter (ADC or A/D). Outputs, one digital with the digital-to-analog converter (DAC or D/A), are needed to drive the actuator.

All control systems, whether microbased or not, need testing at some stage, and so testing methods must be given some attention in a properly designed course on the subject in hand.

Testing may take one of two forms: (1) static, that is, simply checking that a given demand produces a given result, or (2) dynamic, that is how does the system respond to certain changes? Is it sluggish, does it overshoot, or does it go unstable? Testing methods for control systems will vary according to (1) the purpose of the test and (2) the skills possessed by the testers.

Figure 3.18 shows some possible techniques applicable at different stages in the engineering cycle, and as we move from design toward in-service, the equipment and techniques become generally more sophisticated, while the operators become less so. Teaching labs in this field do need to have representative examples of different testing techniques or they are not complete.

3.10   Servomotor Control[b]

The C–1000 Series Buckminster building blocks provide a practical mechanism for controlling the position of a dc servomotor, as shown in Figure 3.19. The C–1600 servomotor control module enables a microcomputer to

---

[b] The material in Section 3.10 is from Application Note 7 and appears here courtesy of the Buckminster Corporation, Somerville, Mass.

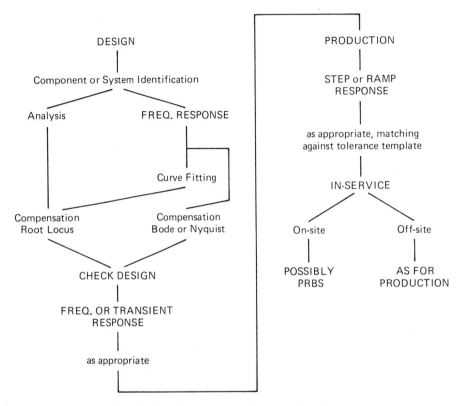

FIGURE 3.18. Testing block procedures for control systems.

directly control the velocity of a dc servomotor. The C–1500 encoder interface module enables a microcomputer to directly measure the position of an optical incremental encoder. The C–1200 power amplifier provides the necessary electrical excitation to the dc servomotor. Figure 3.19 shows a common configuration for closed-loop servomotor control through a microprocessor.

The program first initializes the system by moving the servomotor to home position and then outputting the origin to the C–1500. The program then enters a servo loop, which compares the desired position of the servomotor to the actual position as read in by the C–1500. The position error, which may be a positive or negative quantity, is output to the C–1600. The output of the C–1600 is connected to the C–1200 power amplifier. Motor current and velocity signals are fed back from a current sense resistor and tachometer to the C–1600. The speed of the motor is proportional to the error signal coming from the microcomputer. As the servomotor gets closer to the desired position, the error signal gets smaller, and eventually reaches zero when the motor is in exactly the desired position.

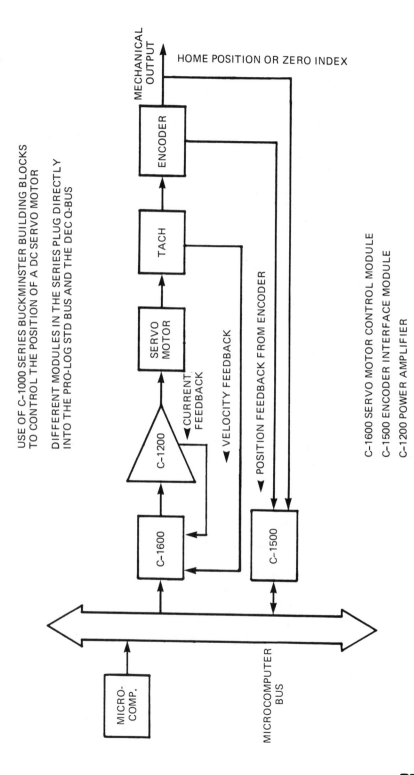

USE OF C-1000 SERIES BUCKMINSTER BUILDING BLOCKS
TO CONTROL THE POSITION OF A DC SERVO MOTOR

DIFFERENT MODULES IN THE SERIES PLUG DIRECTLY
INTO THE PRO-LOG STD BUS AND THE DEC Q-BUS

C-1600 SERVO MOTOR CONTROL MODULE

C-1500 ENCODER INTERFACE MODULE

C-1200 POWER AMPLIFIER

MECHANICAL OUTPUT

HOME POSITION OR ZERO INDEX

ENCODER

TACH

SERVO MOTOR

C-1200

CURRENT FEEDBACK

VELOCITY FEEDBACK

POSITION FEEDBACK FROM ENCODER

C-1600

C-1500

MICRO-COMP.

MICROCOMPUTER BUS

FIGURE 3.19. A servo control system. Courtesy of Buckminster Corp., Somerville, Mass.

Because the velocity feedback is not part of the microprocessor timing, but rather an analog path provided for each motor–tachometer pair, servo-loop stability is maintained at low-frequency updates from the microcomputer.

It is possible for a microcomputer to control eight axes simultaneously and still have time left over for other background functions in the microcomputer.

3.11   Servomotor Control Module[c]

The Buckminster Corporation C–1610 shown in Figure 3.20 is a dc servo pre-amplifier, built on a Pro-Log STD Bus-compatible printed circuit card. The card plugs directly into the STD bus and enables a microcomputer to accurately control the velocity of a dc servomotor. Applications include high performance position and velocity servo systems in laboratory instruments, XY tables, robots, and limited-degree-of-freedom manipulators.

The features of the C–1610 include the following:

- Direct microcomputer control of dc servomotors.
- Two fully independent axes (X and Y).
- 512 velocity levels—8 bits forward, 8 bits reverse.
- Potentiometer-adjustable linear speed range of 256 to 1.
- Potentiometer-adjustable log speed range of 20,000 to 1.
- Potentiometer-adjustable voltage and current limits.
- Separate potentiometer adjustments for lead and lag servo compensation networks.
- Pro-Log STD Bus compatible.
- Optional watchdog timer prevents uncontrolled motion.

The microcomputer can specify the desired velocity of a dc servomotor at any time by outputting an 8-bit speed command to the C–1610 via the STD bus. The C–1610 is designed to utilize two feedback signals to control the velocity of the motor. The first signal is produced by a shunt connected electrically in series with the motor. The shunt develops a voltage that is directly proportional to the current flowing through the motor and corresponds to the torque produced by the motor. The second signal is produced by a tachometer (dc generator) coupled mechanically to the motor. The tachometer develops a

---

[c]The material in Section 3.11 appears here courtesy of Buckminster Corporation, Somerville, Mass.

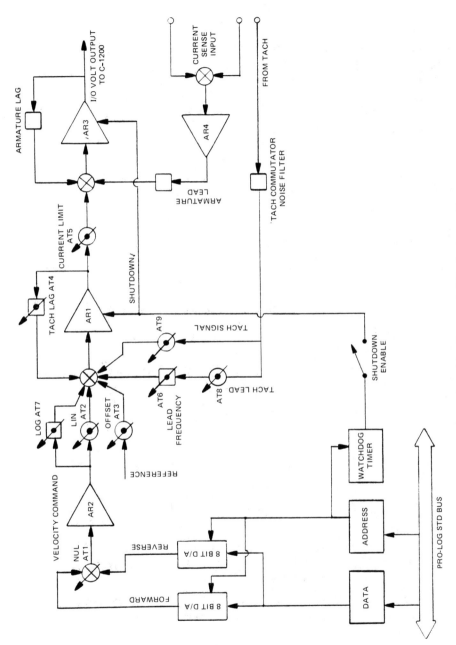

FIGURE 3.20. The C–1610 servo control module. Courtesy of Buckminster Corp., Somerville, Mass.

voltage that is directly proportional to its velocity and corresponds to the velocity of the motor.

The C–1610 outputs a ± 10-V error signal, which is amplified by a C–1200 series power amplifier connected to the motor. The C–1200 series power amplifiers are precision linear amplifiers designed to accommodate a wide range of motor types and sizes.

The C–1610 incorporates a watchdog timer, which prevents uncontrolled motion if the microcomputer stops running. The timer is reset every time the C–1610 is referenced by the microcomputer. If the microcomputer fails to reference the module periodically, the C–1610 will shut down all motion after a short delay. The watchdog timer may be optionally disabled, in which case the C–1610 will not shut down, and the motor will continue moving at the velocity last specified by the microcomputer.

3.12   Review Questions

1. Discuss the meaning of a feedback control system.
2. Discuss the meaning and give examples of open- and closed-loop systems.
3. Draw a block diagram of a closed-loop feedback system with a robot in the system.
4. Discuss the rationale for compensation in a feedback control system.
5. Discuss nonlinearity in a feedback control system.
6. Why are microprocessors used in a feedback control system?
7. Discuss the necessity for the controller of a feedback system.
8. Discuss a servomotor control system.
9. If the forward transfer function G is -1, what is the transfer function, e_o/e_{in} in Figure 3.1?
10. Discuss the Bode diagrams, S-plane plot, and the Nyquist plot.

3.13   Bibliography

Pericles Emanuel and Edward Leff, *Introduction to Feedback Control Theory Systems*, McGraw-Hill Book Co., New York, 1979.

Paul Ford, "Electronics in Energy Control," *New Electronics*, Jan. 13, 1981, pp. 41, 48, 51, 52.

Anthony C. McDonald and Harold Lowe, *Feedback and Control Systems*, Reston Publishing Co., Reston, Va., 1981.

C. J. Savant, *Basic Feedback Control System Design*, McGraw-Hill Book Co., New York, 1958.

4

The Actuators
of a Robot

4.1   Introduction[a]

An *actuator* is a motor, cylinder, or other mechanism that converts one form
of energy to another. In robotics, the resultant action is the mechanical motion, linear or angular, via the robot arm that provides the power to move
each axis. The actuator causes the motion of the robot. At the present time,
several different types are used to power robots.

Electric devices represent about 50% of all the actuators available today.
They include stepper motors, dc servomotors, and pancake motors (discussed
in Section 4.6). Small and big robots also use electrohydraulic actuators,
which comprise about 35% of all actuators used in robots. Hydraulic actuators provide large power-to-space ratios. The simplest type of actuator is the
pneumatic actuator used in about 15% of robots, for example, in the pick-
and-place robots. Grippers and end-effectors (discussed in Chapter 6) also use
pneumatic actuators.

Generally, all robot actuators have advantages and disadvantages; these
are listed in Figures 4.1, 4.2, and 4.3. The basic problems in all actuation sys-

[a]The material in Sections 4.1 and 4.3, in part, is used courtesy of Feedback, Inc., Berkeley
Heights, N.J.

89

Electric Actuators

Advantages

1. Electric actuators are fast and accurate.
2. It is possible to apply sophisticated control techniques.
3. Easily available and relatively inexpensive.
4. Simple to use.

Disadvantages

1. Require gear trains or the like for transmission of power.
2. Gear backlash limits precision.
3. Electric arcing might be a problem.
4. Power limit.

FIGURE 4.1. Advantages and disadvantages of electric actuators.

Hydraulic Actuators

Advantages

1. Large lift capacity.
2. Moderate speed.
3. Oil is incompressible; hence, once positioned, the joints can be held motionless.
4. Offers accurate control.

Disadvantages

1. Hydraulic systems are expensive.
2. They pollute the workspace with fluids and noise.
3. Not suitable for really high speed cycling.

FIGURE 4.2. Advantages and disadvantages of electrohydraulic actuators.

Pneumatic Actuators

Advantages

1. Relatively inexpensive.
2. High speed.
3. They do not pollute the workspace with fluids.
4. Can be used in laboratory work.

Disadvantages

1. The compressibility of air limits their accuracy.
2. Noise pollution still exists.
3. Leakage of air is a major concern.
4. Additional air filtering system and drying system are needed.
5. Increased maintenance and construction requirements.

FIGURE 4.3. Advantages and disadvantages of pneumatic actuators.

tems are (1) power amplification and (2) power control. Figure 4.4 shows a method of achieving these using digital-to-analog conversion (DAC or D/A), electronic amplification, and finally the actuator itself. This system would cope with a linear range of control quite satisfactorily, but unfortunately linear power control is frequently a wasteful business (see Figure 4.5), whereas an on/off control is not. Thus, a dc servomotor, for example, may be more economically driven and controlled by a pulse-width-modulated (PWM) on/off signal. Although power amplification will still be required, there is negligible dissipation and, therefore, waste of power in the drive circuit, and the processor can readily produce a single binary output with PWM applied (see Figures 4.6 and 4.7).

Where very large powers must be controlled, it will usually be necessary to draw the power supply from the ac line, and in this case the modern method would be to use the processor output to drive an appropriate thyristor or triac control unit for a dc or ac motor, respectively.

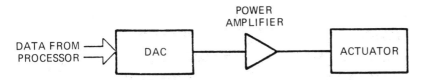

FIGURE 4.4. Basic actuation method.

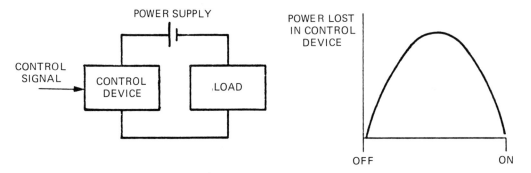

FIGURE 4.5. Power loss in linear control of an actuator.

The choice of an actuator for a given application will mainly depend on its suitability for the purpose. Some factors influencing this are as follows:

1. In factories, air and oil pressure are often readily available, so for economic actuation pneumatic or hydraulic devices may be favored.
2. Electrical actuators are excellent where high accuracy, repeatability, and quiet operation are needed.
3. Pneumatic actuators are generally cheap, clean, and safe and can produce moderately high forces, but the compressibility of air reduces the "stiffness" of pneumatic systems, and this can be a disadvantage in some circumstances.
4. Hydraulic actuators are very stiff in action and can produce very high forces, but they are basically more messy and relatively expensive. Stiffness indicates that the actuator can be very fast acting, and so hydraulic actuators tend to be chosen where speed is essential.

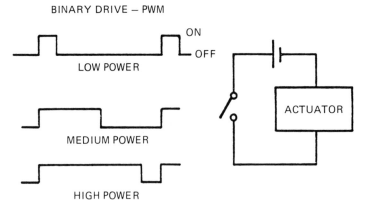

FIGURE 4.6. The pulse-width-method [PWM] control of power.

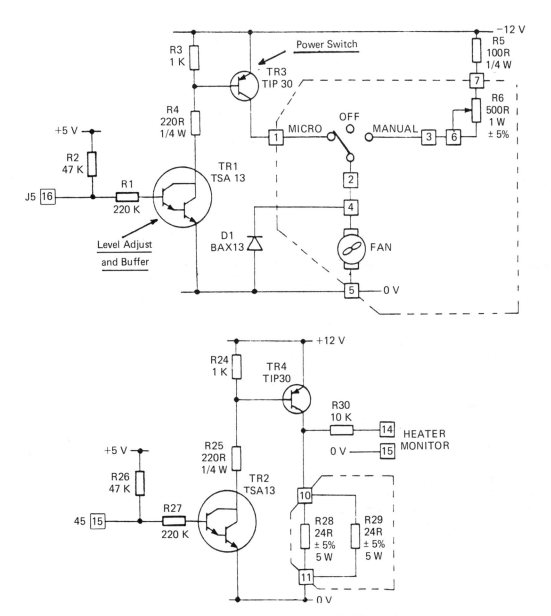

FIGURE 4.7. Schematic for a pulse-width-method control of fan and heater.

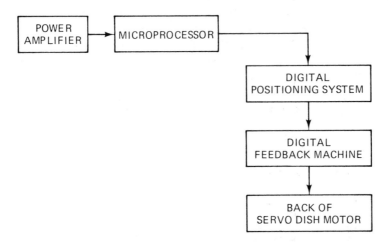

FIGURE 4.8. A simple block diagram of the general actuation system.

A simple block diagram of the general actuation system using microprocessors is shown in Figure 4.8. Note that the actuator is moved by a driven
electronic system controlled by the microprocessor. This is one practical approach to a controller problem. In Figure 4.9 a servo disk motor block diagram is shown using a microprocessor.

4.2   The Control System[b]

Servovalves (an energy input control mechanism of the actuator of hydraulic
or pneumatic servo robots) and circuits (a current control mechanism of the
actuator of electric servo robots) are special energy controllers with feedback
devices. These devices include sensors and related transmitting devices on
both sides of the feedback loop; they send signals to or receive signals from
the controller, closing the feedback loop.

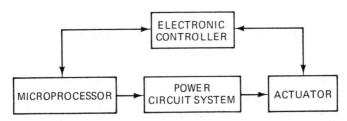

FIGURE 4.9. Block diagram of a system using a servo disk motor.

[b] Section 4.2 is used courtesy of L. F. Rothschild, Unterberg, Towbin, New York City.

The control systems of servo robots include those elements that interface with the controller and the manipulator. Such systems consist of two parts:

1. The control component is the software that initiates, controls, directs, and ends each movement.
2. Within the control component is the feedback system, a communications loop that measures movement and interfaces with the feedback devices of the servomechanisms and the controller. (See Chapter 3.)

Each servomechanism has two sensors (one receiving signals from and the other sending signals to the controller) that create a closed-loop feedback system. The control system, as a whole, uses data generated by the feedback system to monitor movement. The computer then performs calculations to determine the next motion and directs the servomechanisms to move accordingly. (See Chapter 5 for a detailed discussion of robotic transducers.)

Subsequent to program initiation, the computer reads the control command and separates the command into position (and/or rate) instructions for each axis. For point-to-point movement, this command consists of successive end positions (and/or rates). For the more sophisticated, continuous-path robots, the command is effectively a multitude of incremental points (and/or rates) along a path. For each step, the controller calculates the amount of energy required to move each axis/joint to the target position. The computer then sends out sequential control commands to the robot's control system. Each control command is converted into an electrical signal with position (and/or energy flow rate) instructions (i.e., each electric signal has a value proportional to the required energy level necessary to cause movement and maintain path control). These electric signals are sent through each joint/servomechanism to each axis on an ongoing basis to effect movement so that all axes move in a coordinated motion.

Within the servomechanism, the electric signal is the input to the transducer, usually either as an analog signal (for hydraulic motors) or digital signal (for pneumatic or electrical motors). However, technology is changing. Today, the output is often a numerical analog signal for hydraulic actuators and electrical actuators and an on/off signal for pneumatics. This analog output signal then flows directly (or subsequent to amplification) to an actuator, which converts the energy flow—either hydraulic, pneumatic, or electrical—to effect motion.

With hydraulic motors, a sensor on one side of the feedback loop receives analog signals (from the transducer) to either open that valve for a given duration to a given degree or keep the valve closed. When the valve is open, a given amount of a special, noncompressible hydraulic fluid flows from the hydraulic fluid supply through servovalves supplying that motor, which actually moves an element of the manipulator. When the motion is completed, the second sensor, located on the other side of the loop, signals the controller and the valve is closed and the joint locked. With pneumatic motors, the signals are digital, so each valve is either fully opened or closed; when open, the valve allows a burst of air to enter that actuator. Electric motors can also respond to digital signals (relaying current to the motor) and may require an electric brake to lock each joint after the manipulator segment has reached the desired position.

A robot's feedback system must allow for mechanical inaccuracies—weight and inertia factors of payload and the fact that varying amounts of power are required to move the same distance when the directions are different. The feedback loops, using digital control, are implemented either with software or digital hardware. Software, the more flexible alternative, often is used when hardware servos are not available or are impractical. The feedback system works as follows: A sensor on one side of the closed loop determines the position of the links/joints and/or the speed of motion (with encoders, potentiometers, and/or resolvers measuring position, and tachometers measuring velocity). After a motion is completed, the second sensor sends a signal to the computer control to close the feedback loop, and the controller assures that the joints remain in the desired position in sequence (i.e., valves set or electric brakes in place). The controller then performs sophisticated analysis on data received from each second sensor to (1) determine the actual coordinate positioning of the end-effector, (2) compare this position with the respective spatial coordinates stored in memory (the target position), and (3) calculate the difference or error. The computer next determines the energy flow required to move each axis so that the end-effector achieves the target position. Finally, the controller sends out a new set of electric ("error") signals to each servomechanism, and these signals are broken down for each axis.

This process, one of incremental comparative movements, is repeated with great rapidity, so the motion between any two points is rapid but controlled, creating a virtually smooth motion. More specifically, the controller sends input, for a desired point or path, to a sophisticated servomechanism. The controller then attempts to match that target, driving the *load object* (arm, plus any tool) accordingly. As the load lags behind or overshoots the target position, the error signal indicates the magnitude of movement required to compensate, and a command is issued to advance or retard the progress of the load. For point-to-point robots, as the end-effector moves closer to the target position, the error signal for each subsequent step should reduce to zero, creating linear movements. For continuous-path robots, the error message either reduces to zero for linear movements, or the signal achieves a designated near-zero value, which effectively allows the end-effector to round a curve. In the case of both types of movements, once the error signal reaches zero (or near zero for a curve), the controller locks each joint so the axes stop at the correct position, placing the end-effector at the target position. Then the new signals/control commands are sent out, and the next step in the program is initiated.

Intrinsic to this closed-loop feedback system are the timing of measurements, the feedback control computations, and coordination with the environment. Accordingly, software combined with a high degree of engineering skill is required to achieve positional accuracy and minimize oscillation.

The robot's computer may provide a signal to ancillary equipment (that a task is completed), such as an automatic press or a parts-feeding mechanism, for example, a conveyor. With more sophisticated systems, the computer generates timing pauses and controls the velocity and/or speed (acceleration/deceleration) of the manipulator. All servos, regardless of sophistication, have an information flow loop; this loop runs from the controller to the motor, to the moving element, to the sensors, and back to the controller.

4.3 Electrical Methods of Producing Controlled Mechanical Motion

Most electrically driven systems use dc motors with a servo-controlled pulse-width-modulated supply voltage; some low cost robots use stepping motors. With the latter, the motor advances by a fixed increment for each pulse delivered to it. At first sight it would appear that a system could be controlled simply by delivering predetermined numbers of pulses to the stepping motors. This technique has been used on some designs offered to the low end of the market. Repeatability cannot be relied on, since acceleration and load conditions can result in the stepping motor not responding to all the pulses delivered to it. In industrial machines, sensing of actual position is almost always carried out, thus mandating the use of a closed-loop dc servo system.

Electric motors produce rotary motion. To convert to linear motion, a lead screw, usually of the "ball screw" variety, is used (see Figure 4.10). These can be driven directly off a stepper motor, but for rotary motion a gearbox is required. However, ratios in excess of 50 : 1 are frequently required, and a conventional gearbox of this ratio gives substantial problems with friction and backlash. Hence the wide adoption of the harmonic drive, where the reduc-

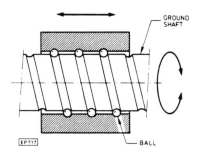

FIGURE 4.10. Ball screw for converting rotary into linear motion.

tion is determined by the ratio of the number of teeth of the larger of two toothed components *to* the *difference* in the number of teeth on the components, instead of it being the ratio of number of teeth on the larger to the number of teeth on the smaller, as in conventional drives. With these units, reductions of up to 320 : 1 can be obtained in one stage. See Figure 4.11 and Chapter 6 for a more detailed discussion of these systems.

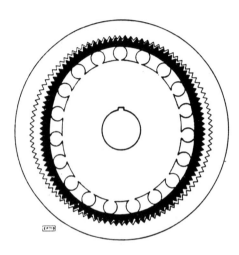

FIGURE 4.11. Harmonic drive gears. The elliptical center revolves, deforming the inner toothed component into contact with the outer toothed component.

Prab Robots, Inc., recently added electric drives to its Model E, FA, FB and FC Robot wrist axes as an option for applications where hydrualic oils should not be used. Electric drive designs are easily incorporated into Prab's robots. The electric cables follow the normal hydraulic routing to the end of arm axes and hand tooling.

Electric drives have not been used on the robot's major axes because, in order to meet the torque-value requirement, the size of the drives would be excessive. Typically, Prab prefers minimizing the robot's spatial requirement in installations, with the robot using only as much floor space as a human worker would occupy.

Prab's Model E and F series robots can handle payloads ranging from just ounces to 2000 pounds. Offering up to seven axes of motion, these robots are used for applications like material handling, machine loading and unloading, investment casting, spot welding, forging, stacking and destacking, and palletizing.

4.4   Alternating-Current Servomotors[c]

The standard servomotor is a two-phase reversible ac induction motor having two windings in the stator (i.e., a fixed or reference phase winding spaced 90° apart electrically from a variable or control phase winding) and a squirrel cage rotor. In operation, the fixed phase is excited at rated voltage, and a second voltage, 90° out of phase, is applied to the control phase. A rotating field results in the stator and generates a rotor torque proportional to the magnitude of control phase voltage. A high torque-to-inertia ratio and a straight-line speed–torque curve are inherent characteristics of this type of motor.

Classes of ac servomotors include the following:

1. *Standard low-intertia servomotors* for precision servos where rapid reversal and high velocity are necessary.
2. *Viscous damped servomotors* that provide greater damping than is available through the technique of reducing the no-load speed. This is normally accomplished by means of a low-inertia drag cup coupling a fixed magnetic field. These motors are normally used in simple instrument servos where high damping and/or low time constants are required, and they can be provided with almost any degree of damping desired.
3. *Inertial damped servomotors* introduce the equivalent of viscous friction into a loop for stabilization without causing velocity lag. Damping is provided on an acceleration or deceleration basis by the viscous coupling of a low-intertia drag cup to a freely rotating permanent magnet flywheel. These motors make possible system cutoff frequencies up to 28 Hz.

This material is presented as an introduction to ac servomotors, since at the present time, few robot manufacturers have adapted ac servomotors.

4.5   Stepper Motors

Stepper motors are electromagnetic devices transforming electrical inputs to mechanical form in discrete and equal increments of rotary shaft motion called steps. A drive circuit provides the usual motor excitation source and steps the motor in response to signal pulses applied to the circuit. A one-to-one correspondence exists between motor steps and these pulses. Net shaft rota-

---

[c] The material in Sections 4.4 and 4.5, in part, is used courtesy of Singer-Kerfott.

tion is controlled by the number of pulses applied to clockwise and counterclockwise command drive circuit terminals, and shaft velocity is controlled by following a constant or varying series of pulses. This synchronous feature makes the motor an ideal integrator and also permits operation of several motors in step with each other at remote locations if all are driven from the same circuit.

Variable-reluctance and permanent-magnet stepper motors are the two most common types. A third type, hybrid in design, contains features of both magnet and reluctance motors. While these types are distinguished principally by rotor construction, stepper motors are also classified by their stator winding arrangements. Most common are 2, 3, and 4 winding motors. Typical step angles for permanent magnet types are 90° and 45°. Smaller step angles are attainable in variable-reluctance and hybrid designs. Step angles of 15° and $7\frac{1}{2}$° are common in these types.

The following are some of the basic terms used in describing stepper motor operation and performance characteristics. The expression *phasing sequence* describes the electrical input required by a particular motor. It is a programmed sequence of connections and polarities for applying dc power to one or more motor windings. Most conventional motors are one of the following six phasing sequences.

Stepper motors can use controllers and single-chip microcomputers to execute a numerical control system for control of stepper motors in industrial robotics. Stored program devices can operate in three modes of operation:

1. Immediate command execution
2. Program entry
3. Stored program execution

Stepper motors have the following features:

1. They operate by electrical pulses.
2. They can "step" forward or reverse a small, but specific, amount with each applied pulse.
3. Continuous pulse signals make them rotate continuously (slewing).
4. Two or more stepping motors can be synchronized to have the same speed.
5. They can be used to position accurately.
6. They are easy to interface with digital control circuits.
7. They can control digital cable drives. Slippage of one step on the exact location on the robot can be controlled by the homing sensor. The homing sensor resets the robot cable to the initial position.

Figures 4.12 and 4.13 show the construction and operation of a stepping motor.

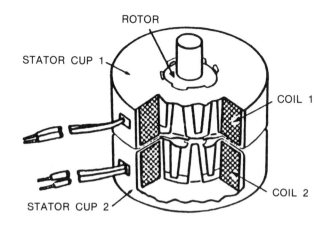

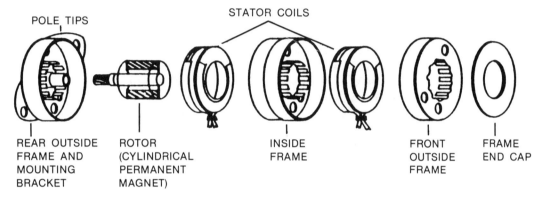

FIGURE 4.12. Construction of a stepper motor.

NUMBER OF STEPS/REVOLUTION: 4–200
MAXIMUM TURNING TORQUE: 0.17–100 OZ–IN
STALL TORQUE: 0.3–200 OZ–IN

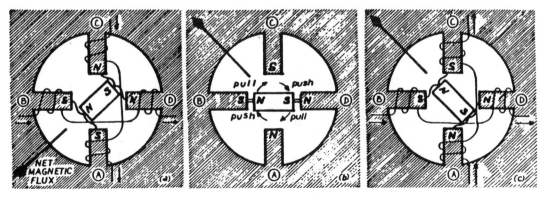

FIGURE 4.13. Working of a stepper motor.

4.6   Direct-Current Servomotors

Figure 4.14 shows a high-performance dc servomotor using an ironless, disk armature (Servodisc™) available from PMI Motors of Syosset, New York, for industrial robotic application. The motors have zero cogging, providing smooth, constant torque even at low speeds for excellent positioning accuracy. Such accuracy is especially important to assure good "repeatability" of the robot arm. Even at high speeds, the motors have high peak torque along with very low inertia. This allows for high acceleration rates and rapid reversing of motion of the robot arm.

The PMI Servodisc motors range in continuous-duty output from $\frac{1}{18}$ to 10 horsepower and will handle radial loads up to 200 lb. Motor speed is 0 to 3000 rpm. They typically weigh less than conventional iron-core motors, providing an excellent power-to-weight ratio. The low axial profile and compact design of the motors mean they can be easily mounted and will not waste space.

The modular design and second shaft extension of dc servomotors allow simple mounting of any type of feedback device. An optional seal oil-proofs the motor for gearbox mounting and provides added protection in harsh, industrial environments. The motors are also available with optional optical encoders and electric brake. Table 4.1 shows the characteristics of the latest JR–1 Servodisk™ dc motor of PMI Motors.

In conjunction with dc servomotors, servo amplifiers are necessary. The switching servo amplifiers, SSA–40, SSA–54, and SSA–75 of PMI Motors features the following:

1. Four quadrants
2. Automatic peak current limiting
3. Wide bandwidth
4. Adjustable compensation
5. Reference voltage outputs of $+15$ V and -15 V
6. Adjustable peak and continuous current limiting
7. Extensive fault protection
8. Spare differential amplifiers provided for signal buffering, inversion, or common-mode elimination

Applications for the servo amplifiers include increment motion control, velocity loops, and positioning systems.

4.7   Electrohydraulic Methods of Producing Controlled Mechanical Motion[d]

Hydraulic systems are widely used on account of their ability to transfer substantial power to a moving part where the weight of an electric motor and

---

[d] The material in Section 4.7 is used courtesy of Feedback, Inc., Berkeley Heights, N.J.

FIGURE 4.14. PMI motion control components for industrial robots. Courtesy of PMI Motors, Syosset, N.Y.

TABLE 4.1
JR SERIES SERVODISC™ DC MOTOR CHARACTERISTICS

1.0	**MOTOR PERFORMANCE: INCREMENTAL MOTION CONTROL**	**SYMBOL**	**UNITS**	**JR12M4C**
1.1	PEAK TORQUE (1)	TP	OZ-IN	1218.8
1.2	CONTINUOUS STALL TORQUE	TS	OZ-IN	109.3
1.3	PEAK CURRENT (1)	IP	AMPS	80.1
1.4	CONTINUOUS STALL CURRENT	IS	AMPS	7.54
1.5	PEAK ACCELERATION WITHOUT LOAD (1) (5)	AP	KRAD/SEC/SEC	71.7
1.6	COGGING TORQUE	TC	OZ-IN	0
2.0	**MOTOR PERFORMANCE: RATED (2)**			
2.1	HORSEPOWER	HP	HP	.36
2.2	TORQUE	T	LBS-IN	7.62
	TORQUE	T	OZ-IN	121.9
2.3	SPEED (8)	N	RPM	3000.0
2.4	POWER OUTPUT	P	WATTS	270.2
2.5	TERMINAL VOLTAGE	E	VOLTS	44.5
2.6	CURRENT	I	AMPS	8.62
2.7	NO LOAD SPEED AT RATED VOLTAGE	NM	RPM	3755.8
2.8	MAX PERMISSIBLE DISSIPATION AT RATED SPEED (3)	PL	WATTS	113.6
3.0	**MOTOR CONSTANTS: INTRINSIC (AT 25 DEG C)**			
3.1	TORQUE CONSTANT	KT	OZ-IN/AMP	15.29
3.2	BACK EMF CONSTANT	KE	VOLTS/KRPM	11.30
3.3	TERMINAL RESISTANCE (7)	RT	OHMS	0.950
3.3.1	ARMATURE RESISTANCE	RA	OHMS	0.730
3.4	AVERAGE FRICTION TORQUE	TF	OZ-IN	6.0
3.5	VISCOUS DAMPING CONSTANT	KD	OZ-IN/KRPM	1.65
3.6	MOMENT OF INERTIA	JM	OZ-IN SEC-SEC	0.017
3.7	ARMATURE INDUCTANCE	L	MICRO HENRY	<100.0
3.8	TEMPERATURE COEFF OF KE	C	%/DEG C RISE	−0.02
3.9	NUMBER OF COMMUTATOR BARS	Z		141
3.10	NUMBER OF POLES OF MAGNETIC FIELD	PF		8
4.0	**MOTOR CONSTANTS: DERIVED (AT 25 DEG C)**			
4.1	MECHANICAL TIME CONSTANT WITHOUT LOAD	TM	MILLISEC	7.47
4.2	ELECTRICAL TIME CONSTANT	TE	MILLISEC	0.14
4.3	SPEED REGULATION AT CONSTANT TERM VOLTAGE	RM	RPM/OZ-IN	4.20
5.0	**THERMAL RESISTANCE**			
5.1	MOUNTED ON ALUM HEAT SINK (8"x16"x⅜")			
5.1.1	ARMATURE TO AMBIENT AT STALL	RAA	DEG C/WATT	1.75
5.1.2	ARMATURE TO AMBIENT AT 3000 RPM	RAA	DEG C/WATT	1.10
5.2	FORCED THROUGH-AIR COOLED (6)			
5.2.1	ARM TO AMB WITH AIR FLOW OF .4 LBS/MIN	RAA	DEG C/WATT	0.80
5.2.2	ARM TO AMB WITH AIR FLOW OF .8 LBS/MIN	RAA	DEG C/WATT	0.50
5.2.3	ARM TO AMB WITH AIR FLOW OF 2.0 LBS/MIN	RAA	DEG C/WATT	0.25
6.0	**PHYSICAL CHARACTERISTICS**			
6.1	MOTOR DIAMETER	D	IN	5.50
6.2	MOTOR LENGTH	LG	IN	3.59
6.3	MOTOR WEIGHT	W	LBS	6.10

Courtesy of PMI Motors, Syosset, N.Y.

gearbox would be prohibitive. Cylinders with pistons similar to those designed for pneumatics are employed, but a low viscosity oil is used instead of air. Being incompressible, very firm positioning and smooth travel are obtained. Here, too, the flow of the fluid is controlled by solenoid-operated valves. Pumps for oil are very compact devices, typically consisting of a pair of electric motor-driven gear wheels in a cavity where oil enters on one side, is trapped between the teeth of the gears, and is expelled on the other side of the cavity. With a single pair of gears, pressures well over 100 bars are readily obtained (1 bar = approximately 14.5 psi = approximately 1 atmosphere).

In each of the ARMDRAULIC robots of Feedback, Inc., low-pressure hydraulics were selected as the most suitable way to produce the powerful and controlled movements necessary for a machine that is to be useful.

Electric motors in general, and stepper motors in particular, are expensive. With these hydraulic robots a single electric motor of the low-cost permanent-magnet variety is sufficient to provide power to all the arm and gripper movements by driving a small hydraulic pump. In hydraulic machines where the drive is taken directly from the pistons, gearboxes, with their expense or friction and backlash problems, are not required, giving a further advantage over electric systems.

Continuous positional control, which is extremely difficult with pneumatics but easy with electric and hydraulic machines, was made particularly easy to implement on these robots by suitable choice of materials, making possible a low-cost inductive coupling system monitoring the piston locations.

In each machine the hydraulic system is as in Figure 4.15, with a pump drawing oil from the sump and pumping via a nonreturn valve into a pressure cylinder in the top of which air is being compressed, thereby acting as a reservoir of power when the pump is switched off, which occurs by means of a pressure switch when the required working pressure of 8 bars (120 psi) is reached. For safety, a pressure-release valve operating at 12 bars is included. The symbols represent the solenoid-operated valves that control the flow of the oil into the cylinders. The return of the pistons is by springs and/or gravity, depending on the function of the cylinder. The interconnections are made with small-bore flexible polyethylene and nylon pipes via screw-in fittings. With these fittings, no problems with oil leaks have occurred, and the machines are very clean in use. No special tools are required for assembling the robots.

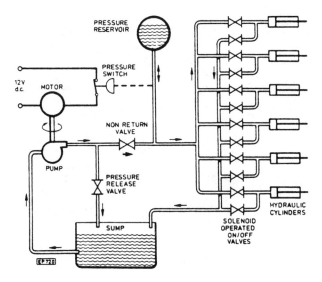

FIGURE 4.15. A hydraulic system.

Some internal arrangements of the ARMDRAULIC hydraulic robot are shown in Figures 4.16 through 4.18. Figure 4.17 shows the internal structure of the ARMDRAULIC robot; it shows the hydraulic accumulator above the bank of the solenoid control valve. Figure 4.18 shows the multipurpose controller. The board shown is configured for use with the ARMDRAULIC CHA 1050 robot.

Figure 4.19 is the internal structure of the ARMDRAULIC robot, showing the "waist" rotation cylinder and solenoid drive electronics.

The Fluid Power Division of Bird-Johnson Company of Walpole, Massachusetts, has developed a series of three modular robotic wrist components (see Figure 4.19). The tightly packed Hyd-ro-wrist is designed to reduce robot design and installation time, while increasing production efficiency.

An assembly of three high-performance Bird-Johnson hydraulic rotary actuators, the 4–1–1 model Hyd-ro-wrist provides three axes of movement: pitch, yaw, and roll. Pitch and yaw axes rotate 180°; the roll axes can rotate a full 280°. The device has one servo valve and feedback transducer for each axis of motion so they can be regulated individually or in combinations.

FIGURE 4.16. Internal structure of the ARMDRAULIC robot, showing the hydraulic accumulator above the bank of the solenoid control valves. Courtesy of Feedback, Inc., Berkeley Heights, N.J.

FIGURE 4.17. Multipurpose controller. Board shown is configured for use with the ARMDRAULIC CHA 1050 robot. Courtesy of Feedback, Inc., Berkeley Heights, N.J.

FIGURE 4.18. Internal structure of the ARMDRAULIC robot showing "waist" rotation cylinder and solenoid drive electronics. Courtesy of Feedback, Inc., Berkeley Heights, N.J.

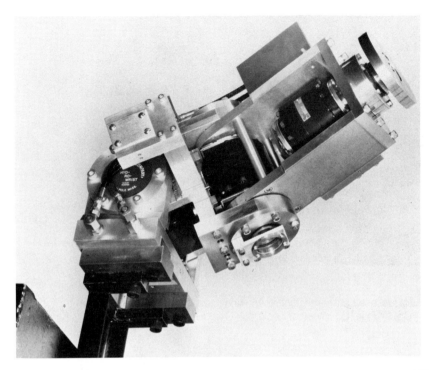

FIGURE 4.19. Modular robotic wrist. Courtesy of the Fluid Power Division of Bird-Johnson of Walpole, Mass.

4.8   Pneumatic Methods of Producing Controlled Mechanical Motion[e]

Pneumatic systems operate by passing compressed air into and out of a cylinder with a piston that provides linear motion (see Figure 4.20). Air is a fluid with very low viscosity and will therefore move through the tubing and cylinder very fast, and cycle times of under a second are realizable. However, gases are compressible; consequently, with a given amount of air in the cylinder the position of the piston will be very dependent on the load being imposed on it. The only way of being sure of positional accuracy is with the use of mechanical stops, with the air providing substantially greater force against the stop than the load to limit the motion of the cylinder. Where rotary motion is required, a lever system, or a rack and pinion, is used. The flow of the air is controlled by solenoid-operated valves. Pneumatic systems require a compressor, which typically is like an internal combusion engine with inlet and outlet valves and a reciprocating piston driven by an electric motor. Being

---

[e] The material in Section 4.8 is used courtesy of Feedback, Inc., Berkeley Heights, N.J.

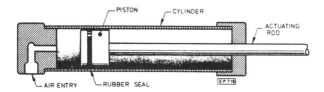

FIGURE 4.20. Pneumatic cylinder.

rather bulky, compressors are not usually integral with robots. Grippers (discussed in Chapter 6) use pneumatic drives because of cost.

Figure 4.21 shows the pneumatic B.A.S.E. robot, Series II–4. The control system consists of a nonservo point to point and three axes with gripper as a robot specification. Resolution is ± 0.001 in.; repeatability is ± 0.010 in.; velocity is one cycle per second per axis. The power section is solenoid operated with four-way valves using 110 V ac.

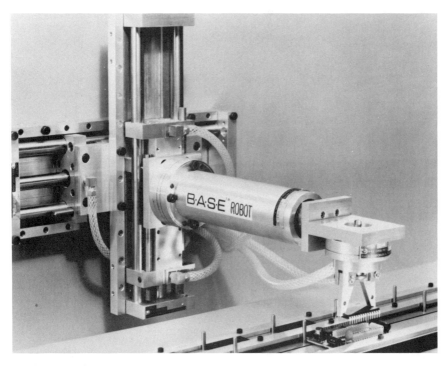

FIGURE 4.21. Pneumatic B.A.S.E. robot, Series II–4. Courtesy of Mack Corp., Flagstaff, Ariz.

4.9 Pneumatic Control System
of the IRI M50 Robot[f]

4.9.1 Pneumatic System

Air is furnished to the IRI M50 Robot (Figure 4.22) at the main manifold in the equipment module and is distributed to the five axes through separate solenoid-operated pressure control valve assemblies and volume chamber/accumulators. The air from each volume chamber/accumulator is directed to the motor/valve assembly for the associated axis. Air from the main manifold is also directed to each axis brake through individual on/off electrical valves.

A pressure regulator, shown in Figure 4.23 is connected to the main manifold and regulates air, which is directed to the pilot pressure manifold. The pilot pressure air is regulated at 80 psi and is supplied to the motor/valve as-

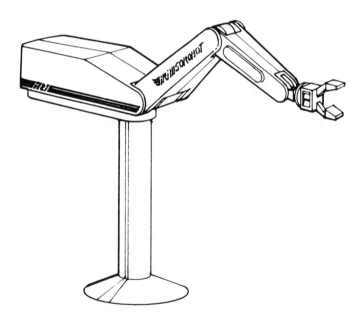

FIGURE 4.22. The IRI M50 robot. Courtesy of International Robomation/ Intelligence, Carlsbad, Calif.,© Jan. 27, 1983.

---

[f] Section 4.9 is taken from IRI M50 Robot Document Reference. Courtesy of International Robomation/Intelligence, Carlsbad, Calif., © Jan. 27, 1983.

SYSTEM BLOCK DIAGRAM

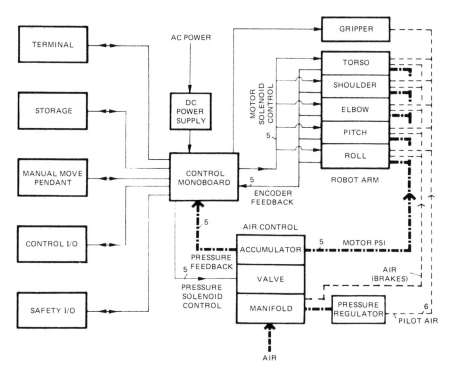

FIGURE 4.23. The system block diagram of the IRI M50 robot showing the location of the pressure regulator. Courtesy of International Robomation/Intelligence, Carlsbad, Calif.,© Jan. 27, 1983.

semblies, furnishing a controlled air pressure for each axis. The spool position in the valve assemblies determines the direction of air flow to the air motors. The spool is positioned as a result of solenoid valve operation. Energizing a solenoid opens a poppet valve, which directs pilot pressure to move a piston inside the valve end, causing pressure to be exerted on the end of the spool valve and moving it to the required position. The air motor direction is reversed by operation of the solenoid on the opposite end of the valve assembly.

4.9.2   Pressure Control

Five pressure control valves are installed in the equipment module, one for each axis. Each control valve assembly contains a solenoid mounted on a valve end and a two-position valve that is operated by pilot pressure when the solenoid is operated. A dithering technique (pulsed air) is used to control the air pressure in the accumulators so that the supply pressure for each motor is constant. The air pressure is modified for different accelerations or for maintaining a velocity.

There are also five volume chamber/accumulators installed in the equipment module, one for each axis. Each volume chamber/accumulator maintains the required air pressure for one axis. The pressure is controlled by the pressure control valve assembly upon which each volume chamber/accumulator is mounted. In addition to serving as a pressure accumulator, the volume chamber/accumulator dampens the air pulses to the air motor. A connection is provided on each chamber for connecting to the pressure transducers.

4.9.3   Air Servo

An air motor is furnished for each axis. The motor for the elbow and shoulder axis delivers up to $1\frac{1}{2}$ hp, while the one for the torso, pitch, and roll axes delivers up to $\frac{1}{2}$ hp. Each motor contains eight vanes and is a reversible type. The vanes are self-sealing and self-adjusting. Exhaust air cools the air motor as it turns, enabling it to be used in temperatures up to 250°F. Lubrication is required and is provided by an air line lubricator in the supply line. Each air motor attaches to its motor stage sprocket assembly by a shaft extending into the sprocket assembly and pinned to it.

The air motor valves are spool type, three-position, in line spring centered, contain five ports, and are operated by solenoid-operated pilot-poppet valves. The ports are numbered for ease of identification: 1, pressure; 2, air motor; 3, exhaust; 4, air motor; and 5, exhaust. The identification plate indicates the ports that are connected when each solenoid is activated. Mufflers are installed on the exhaust ports for noise dampening. A manifold mates each air motor and valve assembly.

4.9.4   Brakes

There are two brake assemblies acting as a caliper on the stage 1 chain sprocket for each axis. All brakes are pneumatically operated except the backup brake for the pitch axis, which is stationary. An electrical control valve is furnished for each axis and directs the air pressure to the brake(s). The pneumatically operated brakes have piston assemblies retained in aluminum alloy housings and contain O-ring seals. All brakes are wire asbestos brake material and are bonded to an aluminum alloy housing. Adjustment is provided for the stationary backup brake used on the pitch axis.

4.10   Review Questions

1. Discuss the advantages and disadvantages of a stepper motor to be used as an industrial robotic actuator.
2. Discuss the use of a servomotor for industrial robotic applications.
3. Discuss hydraulic and pneumatic actuators for industrial robotic applications.
4. Draw a block diagram of a microprocessor-based system terminated by a servo disk motor and a hydraulic cylinder.
5. In any robotic system, when and why are actuators used?

4.11   Bibliography

John J. Pippenger and Tyler G. Hicks, *Industrial Hydraulics*, 3rd edition, Gregg Division, McGraw-Hill Book Co., 1979.

James Sullivan, *Fluid Power*, 2nd edition, Reston Publishing Co., Reston, Va., 1980.

Technical Information for the Engineer (Motors, Motor Generators, Synchros, Resolvers, Electronics and Servo), 10th edition, Singer Company, Kearfott Division, Aerospace and Marine Systems, Little Falls, N.J.

5

Robotic Sensors and Data Acquisition and Conversion Systems

5.1   Introduction

A sensor or transducer is a measuring or control device that changes one type of energy to another. The types of sensors used today in industrial robotics include the following:

1. Parts detection types
2. Tracking sensors
3. Force feedback sensors
4. Robot vision systems
5. Proximity detectors

The robot vision systems are discussed in detail in Chapter 7.

In general, the transducer measures pressure, force, velocity, acceleration, flow, electronic impedance, and many other variables.

Measurement of displacement devices includes the following types of transducers:

   1. Strain gage
   2. Differential type
   3. Capacitive
   4. Inductive
   5. Photoelectric

   In an IBM robotic system, the strain-gage transducer senses the different forces on the gripper. Also important in robotic transducers are the following:

   1. Command interfacing or communication between the robot or teach pendant or whatever the controller is doing.
   2. Sensor interface or the presentation of the transducer output signal to a computer via the RS–232–C or IEEE 488 bus. This is discussed in detail in a subsequent section.

   Tactile sensors respond to contact force that arises between the sensor and a solid object. They include the following:

   1. Microswitch, the most common touch sensor.
   2. Strain gage (discussed in Section 5.4), the most common stress force sensor.

   Proximity switches (inductive and capacitive or photoelectric) are mounted on the robot arm and alert the robot when the object in question is encountered. The robot can pick up the object and move along with the object. Robots may move along a predetermined path with the object. Proximity sensors are discussed in Section 5.6.
   Proximity switches (external to the robot) function in a robot used in searching for an object.
   In this chapter in subsequent sections, resolvers, encoders, and potentiometric transducers for industrial robotics are discussed in detail.
   Optical encoders are used in robot cable drives such as the Rhino^R XR–1, in lieu of stepper motors because they give the exact location on the robot.
   In addition to the potentiometers, resolvers, and optical encoders useful for precise position control in industrial robotics, the following transducers can also be used:

   1. A linear variable differential transformer (LVDT) that changes the position of the armature of the transformer and creates changes in the output voltage.
   2. A capacitive transducer that changes electrode position and alters the capacitance of the circuit.

3. A laser interferometer that changes the position of an object and creates changes in the interference pattern of a laser beam upon it.
4. Other transducers using solid-state technology.

5.2 Force and Torque Measurements

Force and torque measurements are used in industrial robotics. Force is an influence on a body or system producing or tending to produce a change in movement or shape or other effects. Force is a push or pull exerted on a body, while torque is the force that tends to produce a rotation or torsion. Many industrial robotic sensors are available for monitoring force or torque. Force sensors are designed usually for the measurements of a weight or a load.

Force and torque sensors are used to control robot motion, make go/no-go decisions, adjust task and process variables, detect end-effector collisions, determine required actions to unjam the end-effector, coordinate two or more arms, supply compliance (i.e., combined position and force control), and perform high-tolerance tasks with "sloppy" arms.

The FT–1550 is a force/torque sensor system of the Robot Sensor Systems, Inc., of Temple City, California, that performs the preceding tasks in robotics. Basic components of the FT–1550 system include a six degree-of-freedom solid-state piezoresistive force/torque transducer with 4 ft of flexible cable, a preprocessing box, 20 ft of connecting cable, and a high-speed microcomputer processing unit.

The microcomputer processing unit automatically resolves the forces and torques applied to the FT–1550 transducer into six equivalent cartesian force/torque components and transmits the results from the robot at better than 100-Hz rates.

The processing unit contains an RS–232–C serial port with adjustable baud rate to 19,200 baud and a 16-bit TTL parallel interface with handshake. The serial port can be used to transmit force/torque information to the robot controller or a readout device, such as a cathode-ray tube.

5.3 Strain Gage Measurement

The resistance of a wire is directly proportional to its length and resistivity and inversely proportional to its cross-sectional area. If we stretch a wire, its resistance will change. This simple theory is used as the basis for the class of transducers called strain gages. A strain gage transducer is a displacement device. Strain gages are commercially available and are used in industrial robotics to measure force or torque.

The unbounded strain gage elements use one or more wire filaments of resistance stretched between supporting insulators. The supports are either attached directly to an elastic member used as a sensing element or are fastened independently, with a rigid insulator coupling the elastic member to the taut filaments. The displacement (strain) of the sensing element causes a change in the filament length, with a resulting change in resistance.

The piezoresistive strain gage exhibits more than 100 times the unit resistance change of a foil gage for any given strain. This means that if semiconductor gages are connected as the arms of a Wheatstone bridge, a very large output voltage can be produced, eliminating the need for subsequent amplification. However, such large resistance changes produce large unbalances in a Wheatstone bridge with constant-voltage excitation, resulting in very nonlinear outputs. This problem can be solved by exciting the bridge from a constant-current supply.

Actually, the resistance change of a semiconductor strain gage as a function of strain is not completely linear over its total strain range. This results in nonlinearity in some transducers, even with constant-current excitation. Foil gage transducers require amplification because of low bridge output. The linearity of the output signal is usually not a problem.

5.4   Potentiometric Transducers for Industrial Robotics[a]

Linear displacement can be achieved by using a pot (slide bar). The motion of the slider results in a resistance change that can be made linear depending on the way the resistance wire is wound.

The mechanical resistive elements rely upon the resistance variation produced by a mechanical input to its movable slider. Typical construction of the resistance requires precision elements.

Figure 5.1 shows a diagram of the Durapot[R]. Durapot provides a smooth, continuous dc voltage directly proportional to shaft rotation. Each unit contains an electromagnetic rotary transducer with all solid-state excitation and conditioning electronics. Electronics may be supplied within the transducer housing or in a separate enclosure. Three grades of transducer enclosures are available: standard NEMA 1, heavy-duty NEMA 12 and water-resistant NEMA 13.

Features of the Durapot include the following:

- No wipers to wear
- Long life with 100 million rotations minimum
- Deadband less than 0.02°

---

[a] The material in Section 5.4 is used courtesy of Astrosystems, Inc., Lake Success, N.Y.

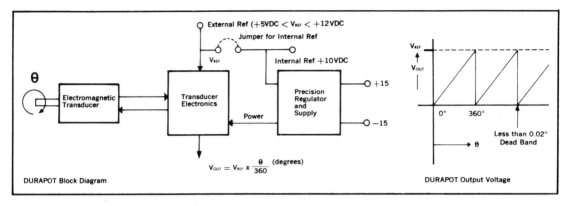

FIGURE 5.1. Block diagram of the Durapot. Courtesy of Astrosystems, Inc., Lake Success, N.Y.

- Continuous rotation
- Short-circuit proof
- 0.05% linearity
- 0.1 oz. in. starting torque
- Infinite resolution
- Up to 2000 Ω load
- Internal or external reference

5.5   Proximity Transducers[b]

Proximity sensors in robotics use inductive and capacitive circuits. (Photo-electric devices are discussed in Section 5.5.1 and Section 6.13.)

The *mechanical command sender* is defined as a limit switch, position switch operating by mechanical sensing. The detection elements are levers, rolls, tappets, and so on, whose angle of approach, path of approach, and slowdown will have to be calculated. The actuation of mechanical command senders requires mechanical force. The service life of the switching contacts is limited; as is the switching frequency. The signals can only be used to a limited extent for further processing in the modern, contactless (digital) controls.

The *electronic command sender* is defined as an inductive and capacitive proximity switch with contactless sensing and switching output. Proximity is converted into switching commands. No mechanical force is required. The

[b] The material in Section 5.5 is from the *8000 Series Technical Product Information* and is reprinted courtesy of the Automatic Timing and Controls Company, King of Prussia, Pa.

movement of the object to be detected is not impeded. One design serves many applications. Electronic command senders have a contactless switching output.

The inductive proximity sensor is based on an *LC* oscillator circuit. A high-frequency, directional, alternating field is set up with an open shell core with a coil. This alternating field is also termed the active zone. If an electrically conductive material such as metal is now introduced into the active zone, the alternating field is weakened. This then leads the system to the breakdown of the oscillation. These two states are: oscillator oscillating (no object in the active zone) and oscillator not oscillating (object present in the active zone). The signal is passed from the oscillator via a high-impedance selection device through a rectifier to a sweep amplifier. Depending on the triggering impulse, the output stage from the sweep amplifier becomes a high resistance or open circuit or a low resistance or closed circuit. Inductive sensing contactlessly detects electrically conductive materials such as metals.

On the other hand, the capacitive proximity sensor uses an *RC* oscillating circuit. In this process, a changeable capacitance is used to trigger the switching operations. The active zone is created by two cup-shaped electrodes creating the capacitance. If an object is introduced into the active zone, the capacitance of the oscillating circuit changes. Here also there are two clearly defined states:

1. No object in the active zone and the amplitude of oscillation is small.
2. Object present in the active zone and the amount of oscillation is large.

These two states, oscillator not oscillating (no object in the active zone) and oscillator oscillating (object present in the active zone), are then passed to the output stage after relevant evaluation by means of a high-impedance selection device, rectification, and sweep amplifier.

The capacitive sensing system detects all metallic and nonmetallic materials that have a high-resistance connection to the earth ground or which themselves possess a certain reference potential.

Proximity sensors are used to give the robot a sense of sight and touch.

5.5.1 Pulsed Infrared Photoelectric Control for Industrial Robotics[c]

Pulsed infrared photoelectric controls are used in industrial robotics for presence sensing any type of object. These devices can be used to detect the product on the infeed or outfeed conveyer to and from the protected robot work station.

---

[c] The material in Section 5.5.1 is used courtesy of Scientific Technology, Inc., Mountain View, Calif.

The STI 3030-Series OMNIPROX™ (Figure 5.2) is a complete solid-state modulated infrared beam detection and control system with modular versatility. As a proximity sensor, it detects anything entering or leaving its field of view, metallic or nonmetallic, liquid surfaces, or gas clouds. Its high adjustable sensitivity permits a choice of setting for detecting the surface or for seeing through transparent and translucent materials, liquids, or clouds. It reads code marks or color changes.

The range of the STI 3030-Series is adjustable up to 12 in. (30.5 cm) in the standard model, and longer-range models are available.

The 3030-Series sensor head is totally sealed and shock proof. It may be mounted anywhere, indoors or out, submerged or in a vacuum, up to 100 ft from the control electronics. It disregards ambient light, atmospheric contamination, and thin-film accumulations of oil, dust, water, and other airborne deposits. Cabling to the sensor is low voltage.

A variety of standard plug-in output and control options adapt the 3030-Series for a broad range of applications. It is used for web break detection, safety controls and intrusion detection, and liquid level or filling level inspection. It is applied for sensing, counting, routing, positioning, inspecting, measuring, and all sorts of other uses, particularly for automated machine control.

Multiple sensors can be placed remotely from a single control card for multipoint monitoring, sequential operation, and other applications where more than one spot or area must be watched.

The STI 7060-Series shown in Figure 5.3 is a complete solid-state transmitted infrared beam system for precise short range (up to 25 ft) detection and control applications. An accessory masking kit limits beam diameter for fine wire, thread, filament, and small parts detection.

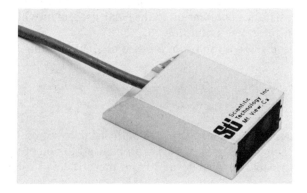

FIGURE 5.2. Photoelectric proximity sensor. Courtesy of Scientific Technology, Inc., Mountain View, Calif.

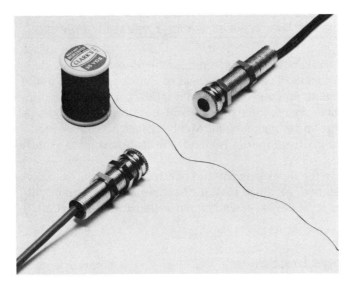

FIGURE 5.3. Photoelectric sensor transmitted beam. Courtesy of Scientific Technology, Inc., Mountain View, Calif.

The small, totally sealed, stainless steel, shock-proof 7060 sensor heads may be mounted anywhere, indoors or out, submerged or in a vacuum, up to 100 ft from the control electronics. The detector disregards ambient light interference, atmospheric contamination, and thin-film accumulations of oil, dust, water, and other airborne deposits, because a form of automatic gain control maintains sensitivity. Cabling to the sensor is low voltage.

With the 7060-Series transmitter and receiver heads mounted side by side, the system will operate in either the retroreflective or proximity modes. The 7060-Series is simple to set up and operate, requiring no focusing. A visible red LED indicator in the control box glows when the sensor heads are properly aligned. It also permits visual monitoring during operation. A potentiometer provides range sensitivity adjustment. Either a "Beam Break" or "Beam Make" operation mode may be selected.

Maintenance is practically never necessary because there are no lamps to burn out or any other components that deteriorate rapidly or periodically.

A variety of standard plug-in control and output options adapt the 7060-Series for a broad range of industrial applications. Major uses include automated machine and other applications where small parts of thin sections must be accurately sensed, counted, routed, positioned, inspected, measured, or controlled.

Multiple sensors can be placed remotely from a single control for multipoint monitoring, sequential operations, and other applications where more than one spot or area must be watched.

5.6 Motion and Angle-sensing Transducers[d]

Motion measurements convert linear and angular displacement and their derivatives to electrical signals. The measurement of shaft angle is one of the most prevalent requirements in control, instrumentation computers, and industrial robotics. Almost every machine process and monitoring system has a rotating shaft somewhere in its mechanism. Mechanisms for converting to shaft rotation to translation (linear motion) extends the usefulness of the shaft-angle sensing.

Shaft angle is used in the measurement and control of position, velocity, and acceleration in one, two, or three spatial dimensions and in many different sets of parameters must be sensed and or controlled and is important in industrial robotics.

5.6.1 Shaft Angle Encoders[e]

There are two basic types of electronic shaft angle encoders called *incremental* and *absolute.* An incremental encoder is a pulse generator whose output describes a change in shaft position as a series of pulses. An absolute encoder measures the actual position of the shaft.

Incremental encoders use up–down counting circuits to accumulate incremental output pulses and thereby indicate finite changes in shaft position as an accumulation of shaft movements.

Incremental encoders are relatively simple to install and operate and have a cost advantage over the more sophisticated encoders. Furthermore, the signal generated by these encoders is basically digital so that it can be easily connected to a computer. For this reason, they have found wide acceptance in machine motion-control systems. However, the low cost of the incremental encoder must be balanced against other important factors in use.

Incremental encoders have several disadvantages when compared to other position measurement sensors. Since encoders tend to be used in electrically noisy environments, a noise pulse can create a false count that will remain unless corrected. A similar problem results if the shaft position is changed during power-down conditions. During the downtime period, the accumulated count will be lost unless a nonvolatile memory or battery backup is used. Self-correction or synchronization is usually accomplished by recycling the shaft through mechanical zero, while resetting the counters to zero.

---

[d] The material in Section 5.6 is used courtesy of the ILC Data Device Corp., Bohemia, N.Y.

[e] Section 5.6.1 is from "Shaft Angle Encoders, AN–1A2" and Section 5.6.2 is from "Absolute Encoder Shaft Coupling, AN–1A4." The material is used courtesy of Astrosystems, Inc., Lake Success, N.Y.

Absolute encoders measure actual shaft position. Both optical and electromagnetic techniques are in use in absolute encoder systems. Absolute encoders are not upset by power-down conditions; therefore, synchronization and resetting are not necessary after a power outage.

An absolute optical encoder uses a shaft-mounted disk having concentric rings of alternate transparent and opaque segments, a light source, and a group of photocells that respond to the alternate blocking and passing of light as the disk rotates. A 10-bit binary encoder uses 10 concentric rings with the outer (high-resolution) ring providing 2^{10} pairs of transparent–opaque segments, and the inner ring divided into one opaque and one transparent semicircle. Thus, the inner ring would indicate whether the shaft was positioned between $0°$ and $180°$ or between $180°$ and $360°$. The second ring would specify the correct quadrant, while each additional ring would further improve the resolution by a factor of 2. The outer (tenth) ring would then define shaft position to 1 part in 2^{10}.

An incremental optical encoder only requires one concentric ring divided into as many segments as resolution requires. The output is a series of pulses generated as the disk rotates. Electronic circuitry in close coupling with the optics amplifies and shapes the pulses, which then go to an external up–down counter.

5.6.2 Absolute Encoder Shaft Coupling

A mechanical coupling is required to connect the absolute encoder transducer shaft to the machine shaft that is to be monitored. This coupling must be a high-quality device with predictable and repeatable motion. Any hysteresis or play in the coupling can introduce unacceptable error into the system. The coupling may also have to provide mechanical scaling between the machine shaft and the transducer to make the encoder output read in appropriate units.

The coupling must be able to compensate for any misalignment of the two shafts. When the machine shaft is subject to heavy shock vibration, it is recommended that an intermediate shaft, supported by its own bearings, be used to isolate and protect the transducer.

The torque requirements of the transducer are so low that the transducer bearings require no maintenance. It is external factors such as improper installation, misalignment, or shock loading that can cause premature bearing failure.

Once a specific absolute encoder has been selected based on total count and resolution requirements of the application, the mechanical interface can be determined. Both rotary and linear motion can be measured and either a single turn or multiturn system can be applied. The scaling discussed next applies to all systems.

For applications where the transducer is to be connected directly to the machine shaft, a 1:1 coupling is appropriate. This coupling can be direct shaft to shaft using a Rembrant or equal coupling, or a 1:1 gear, timing belt, or similar device may be used. A 360 count per turn (cpt), single-turn system is the most common direct-connected system where the output is read directly in degrees.

For all other encoder installations, a ratio coupling system is required. (see Figure 5.4). The ratio system can be a gear train, timing belt, rack and pinion, cable and pulley drive screw, or similar device. Proper determination of the interface ratio requires specific details of the machine motion.

The required ratio is calculated as the desired count per turn (single turn) or full travel (multiturn) divided by the selected encoder count per turn or full travel capability.

$$\text{Ratio} = \frac{\text{count per turn of application}}{\text{count per turn of encoder}} \qquad [5.1]$$

For example, a machine shaft rotates one revolution in driving a full travel distance of 800 mm. To encode 800 units in one revolution, a 1000 count per turn encoder can be used. The ratio required is 800/1000 or 0.8. This means an interface should be provided so that the transducer shaft turns 0.8 of a revolution for each full rotation of the machine shaft. The encoder will then read 800 counts for one revolution of the machine shaft.

The same ratio calculation applies to both single and multiturn systems. For example, if 50 revolutions of a machine shaft equals 36.000 in. of full travel, a 36-turn encoder geared 50:36 to the machine shaft will provide the required encoding to the specified resolution. If only 36.0-in. resolution were required, a 360 count per turn encoder with a 50:1 gear train would be appropriate.

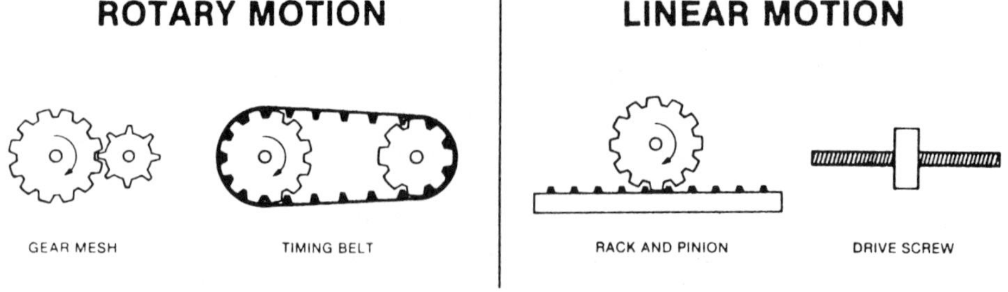

ROTARY MOTION | LINEAR MOTION

GEAR MESH   TIMING BELT | RACK AND PINION   DRIVE SCREW

FIGURE 5.4. Ratio coupling systems. Courtesy of Astrosystems, Inc., Lake Success, N.Y.

5.7   Optical Encoders for Robotic Systems[f]

As an introduction, digitizing of analog linear or rotary information requires use of an encoding position transducer. Linear encoder transducers include the Inductosyn[R], an inductive device, and linear optical encoders. Widely used transducers (position encoders) for robotics are potentiometers (usually for course information), optical encoders, and resolvers. The incremental optical encoders are more commonly used because it is less expensive. For absolute position control, the resolver and resolver-to-digital converter are much less expensive than absolute optical encoders. In some cases, the resolver-based system is equal in price to incremental optical encoders because it has the advantage of absolute positioning in robotics. Robots using resolvers and Inductosyns have up to six position sensors per channel. Potentiometers are used in many cases as a coarse position input in a two-speed system.

Potentiometers are generally the lowest in cost, but face reliability problems, such as wiper wear, and are more susceptible to the environment than the optical encoder or resolver transducers. For resolution and accuracy in the range of 10 to 14 bits, the optical encoder and the resolver are more frequently used than the potentiometer. Let us compare the optical encoder and resolver digital conversion systems.

The basic components of the rotary optical encoder (Figure 5.5) are the coding disk, light source, light detector, and the signal-processing electronics.

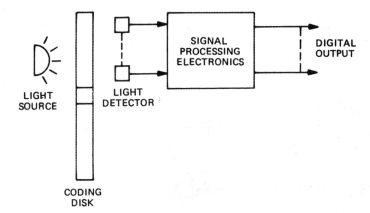

FIGURE 5.5. The four basic components of an optical encoder are the light source (an incandescent lamp or LED), the rotating code disk, the light detector (e.g., phototransistor), and signal-processing electronics. Courtesy of the ILC Data Device Corp., Bohemia, N.Y.

---

[f]The material in Sections 5.7, 5.9, and 5.10 is used courtesy of the ILC Data Device Corp., Bohemia, N.Y., and Measurements and Control, Pittsburgh, Pa.

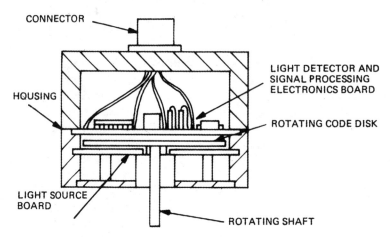

FIGURE 5.6. Construction of the optical encoder. Resolution and accuracy are obtained through the precision of the rotating and stationary mechanical components. Encoders are available with connectors mounted axially or radially. Courtesy of the ILC Data Device Corp., Bohemia, N.Y.

The light source can be either an LED or the incandescent lamp type. LEDs can generally withstand mechanical vibrations that lamps cannot, and are better for rugged applications; but the tradeoff is in its sensitivity to temperature and lower light output. The light detector (e.g., phototransistor) generates an output when excited by light energy; Figure 5.6 shows the mechanical construction. The most important component is the coding disk (Figure 5.7), a mechanical device usually made of glass or plastic. The disk has opaque and transparent areas deposited onto it in a concentric pattern, which determines the accuracy and the resolution of the encoder.

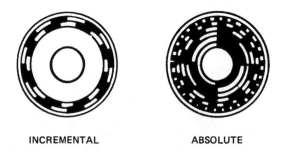

INCREMENTAL   ABSOLUTE

FIGURE 5.7. The shaft position information, absolute or incremental, comes by way of coding disks. The cost of absolute encoders is more than twice that of incremental encoders because of their complexity. The output code format is determined by the physical layout of the code pattern on the disk. Courtesy of the ILC Data Device Corp., Bohemia, N.Y.

Shaft position is determined from the signal pulses that are generated as a result of the coding disk cutting the light beam. The number of opaque and transparent areas increases with increase in resolution, requiring the encoder to be physically larger for higher resolutions. High resolution and accuracy demands great mechanical precision by which the concentric rings of the pattern must be deposited.

Rotary optical encoders are available as *incremental* or *absolute position* types, as discussed in Section 5.2.1. Incremental optical encoders generally have two signal outputs that are in quadrature (i.e., 90° phase difference) for position and direction data, plus a marker signal output for initialization. The absolute optical encoder makes absolute position data available in the form of natural binary, gray code, or binary-coded decimal formats. The code is determined by the coding pattern on the disk, while the format in the resolver system is accomplished by the electronics.

The advantage of the absolute system is that it can maintain position information even during a power failure, whereas in the incremental system the encoder has to be rotated to the marker signal for initialization once the power is back on. In an absolute endoder there are many signal output lines (one for each bit position) going to the digital data processor.

The output of the light detector requires amplification and signal shaping. The signal-processing electronics detect the signal, generate digital pulses, shape the output, and filter and amplify for transmission. A stationary disk between the coding disk and the light detector and a lens to form the light are often used to enhance the quality of the light beam.

5.8   Encoder Interface Module[9]

The Buckminster Corporation's C–1510 is an incremental encoder-to-computer interface, built on a Pro-Log STD Bus-compatible printed circuit card. It enables a microcomputer to keep track of position based on the outputs of incremental shaft encoders. It is designed to interface one or two incremental shaft encoders to the STD bus.

Applications for industrial automation include robotics and XY table control. For example, the module can be used to control the position of servomotors by closing the feedback loop through a microcomputer. C–1510 can also enable a microcomputer to control multiple axes simultaneously.

The C–1510 features 8-bit TTL parallel input to detect zero index from the encoder, limit switches, and fault conditions, and 8-bit TTL parallel output to control brakes, valves, solenoids, and lights. The C–1510 also provides $+5$ V for each encoder.

---

[8] The material in Section 5.8 is used courtesy of Buckminster Corp., Somerville, Mass.

The signal derivation and functional diagram of the C–1510 encoder interface module are shown in Figure 5.8.

The C–1510 derives increment and direction information from two quadrature signals generated by the encoder. Both are binary TTL signals, each alternating from a low state to a high state as the encoder input shaft is rotated or the linear encoder head is moved.

The module can be configured as two 16-bit counters or one 32-bit counter. Each channel has a counter associated with it whose contents may be read or written by the microcomputer. Each counter keeps track of the position information produced by one incremental encoder, increasing or decreasing by one unit whenever the encoder crosses an edge. The phase relationship between the two signals determines the direction of motion.

The contents of the counter thus represent position relative to an origin. The contents of each counter appear on the microcomputer's bus. Position information can be obtained at any time by reading the counter, and the origin can be set to any desired value by writing that value into the counter.

The counters of the C–1510 are internally buffered, which enables the computer to read clean position data at any time, even while the encoders are moving and the counters are changing.

Continuous operation of the C–1510 is inherent in the use of incremental shaft encoders, since they have no dead spots. The range of position measurement is limited by the size of the counter. The accuracy is limited only by the resolution of the encoder.

SIGNAL DERIVATION

FUNCTIONAL DIAGRAM

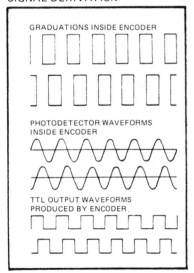

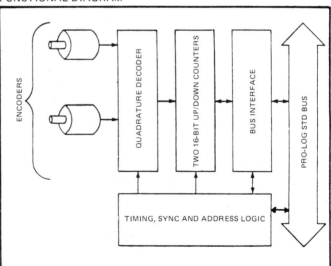

FIGURE 5.8. Signal derivation and functional diagram of the C–1510 encoder interface module. Courtesy of Buckminster Corp., Sommerville, Mass.

5.9   Resolvers

The resolver types of position transducers are primarily transformers with variable coupling as a function of rotor position. The resolver stator contains two windings that are 90° apart (Figures 5.9 and 5.10). Because resolver outputs (sine and cosine) are easier to work with than synchro outputs, industrial users tend to use the resolver transducer.

The resolver is similar to an electric motor in its rugged construction. The stator consists of windings, which are 90° apart, on a slotted cylindrical lamination; the rotor is made up of a shaft, lamination, windings, and slip rings. Recently, brushless resolvers that couple the primary excitation voltage through a transformer, rather than through brushes and slip rings, have become more widely used.

The optical encoder uses optomechanical techniques to directly produce digital wave forms that correspond to shaft position; the resolver system converts shaft position into electrical information through electromagnetic interaction (Figure 5.11) and then converts that analog information into digital data. The primary difference between the optical encoder and the resolver system is division of cost between mechanical and electronics sections; the major cost in an optical encoder is in the mechanical portion, whereas in the resolver system the major cost is in the electronics.

Industrial users require resolution in the 10- to 14-bit range. The increase in cost due to higher resolution is relatively low for resolver systems because the resolver cost remains the same, and the electronics cost increases only about $50 to $70 for every 2-bit increase in resolution. Advances in electronics have made the resolver system competitive with the optical encoder.

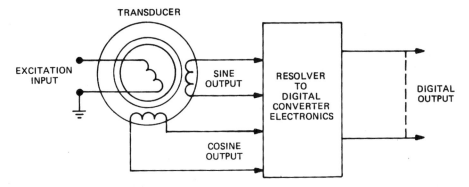

FIGURE 5.9. The resolver to digital encoder (resolver/encoder) system consists of the electromechanical resolver transducer and digitizing electronics. The electronics are available as a discrete module or as a hybrid microelectronic module. Courtesy of the ILC Data Device Corp., Bohemia, N.Y.

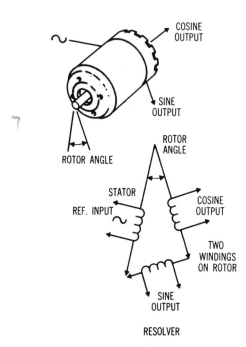

FIGURE 5.10. Resolver transducers provide a great deal of resolution in micro-processor-based rotational controls by measuring phase-angle differences between an ac reference voltage input and output of the rotor coils. Courtesy of the ILC Data Device Corp., Bohemia, N.Y.

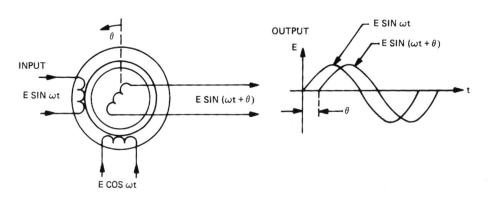

FIGURE 5.11. Rotor position [θ] can be determined by comparing the time phase shifted output signal with the input signal. By counting the number or pulses between the zero crossing of the two signals, rotor position [θ] can be converted into digital format. Courtesy of the ILC Data Device Corp., Bohemia, N.Y.

There are three widely used techniques for converting resolver outputs into digital data: *tracking, successive approximation,* and *time phase shift.* The Type II tracking-loop technique (Figure 5.12) used by DDC uses the modulated sine/cosine outputs of the resolver. Through trigonometric manipulation, an error signal sin $(\theta - \phi)$ is generated, where θ is the angle to be digitized and ϕ the angle generated in the converter. This error signal is fed into the processing circuit, which generates clock pulses as a function of the error voltage level, which is then converted into digital output through an up–down converter. The Type II tracking loop has two integrating stages, one in the error processor and VCO circuitry, and the other at the up–down counter. This type of converter has no velocity error; the data are always fresh.

The *successive-approximation method* (Figure 5.13) uses a sample-and-hold at the front end of the converter to retain the momentary sine and cosine information. The sample is taken at the peak of the reference carrier signal, and the dc signals are forwarded to the input process circuitry for trigonometric manipulation, whereby the error signal sin $(\theta - \phi)$ is generated. A sequence of successive approximations follows which causes the converter to approach the desired angle. The successive approximations start with the most significant bit (MSB) and proceed down to the least significant bit (LSB).

One weakness of this method is the "staleness" error, which results from the fact that a sample is taken only once per carrier period. A multiplexed converter system using the above concept has been developed by DDC. It consists of a central converter module and one or more input processor modules. Since the cost of these converters is higher than the standard tracking converter, the cost tradeoff occurs at approximately six channels. Cost per chan-

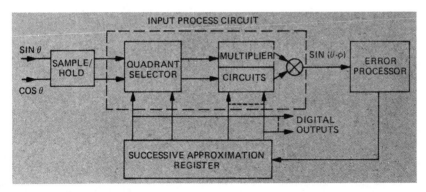

FIGURE 5.12. Type II tracking-loop converter includes two integrating stages, one in the error processor and VOC block, the other in the up–down counter. There is no velocity error and the data are always fresh. Courtesy of the ILC Data Device Corp., Bohemia, N.Y.

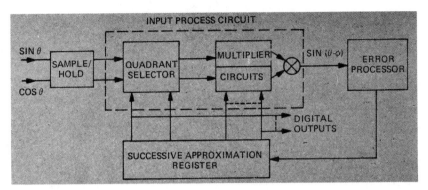

FIGURE 5.13. The input process circuit of the sample-hold/successive-approxima-tion method is similar to the input section of the Type II tracking technique. While the sin $[\theta - \phi]$ in the Type II is a modulated signal, the sin $[\theta - \phi]$ in the above circuit is a dc signal. The up–down counter in the Type II has been replaced with a successive-approximation register. Courtesy of the ILC Data Device Corp., Bohemia, N.Y.

nel for a six-channel 14-bit absolute system is $250 to $300, including the transducer. This multiplexed system is cost effective when the number of channels is large and the system can withstand the "staleness" error.

The methods of conversion described operate on the sine and cosine outputs of the resolver. The time phase-shift method excites the stator with two signals that are 90° apart. The output of the rotor is a signal that is time phase shifted by an amount proportional to the rotor position θ. The input and output signals are compared in time, whereby position θ can be converted into digital data with the use of digital circuitry. The problem with this method is sensitivity to temperature changes.

As mentioned, the basic difference in the position-detection method between the optical encoder and the resolver is that an optical encoder utilizes electro-optomechanical means of generating electronic position information through basic electromagnetic interaction.

5.10   Applications for the Absolute
Shaft Encoders

Although there are linear transducers (such as Inductosyn or linear optical encoders), rotary transducers can be used for both rotary and linear requirements. In rotary position applications, the transducer can be coupled to the moving shaft by the use of flexible coupling, direct gearing, or drive belts.

When gearing is used, the gear ratio can be any practical ratio desired. For example, suppose that the primary shaft turns 100 revolutions to attain the required displacement; however, the application requires only one turn of the transducers (Figure 5.14). This is an example of a single-turn application; the

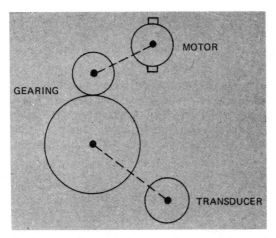

FIGURE 5.14. By the use of gears, the transducer can be made to operate as a single-turn or a multiturn device. If the gear ratio is reversed in the above figure, a higher resolution can be attained. Courtesy of ILC Data Device Corp., Bohemia, N.Y.

absolute system will always give absolute position information. However, when the rotary transducer is attached to a lead screw, it must have the capability of informing the electronic controller that it has gone through a full revolution. This signal, which is usually called a *marker* or a *major carry* signal, is necessary so that the controller knows how much actual displacement took place. For example, if the lead screw pitch is 0.1 in. and the workpiece must traverse 3 in., the equivalent movement of the rotary transducer is 30 revolutions.

For linear-displacement requirements, a rotary transducer can be attached to a lead screw (Figure 5.15). High resolution can be achieved with the proper lead screw pitch. If a lead screw with a pitch of 0.1 in. and a 12-bit (4096-count) position-to-digital conversion scheme is used, the linear displacement resolution is 0.1 in. divided by 4096, equivalent to 24.4 microinches.

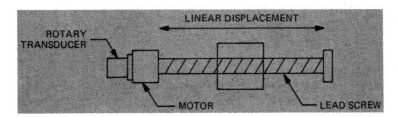

FIGURE 5.15. Linear displacement measured with a rotary transducer. The pitch of the lead screw determines the linear displacement and the rate of displacement. High resolution can be attained; e.g., a pitch of 0.1 in. can be resolved into 25 microinches with a 12-bit (4096-count) position-to-digital encoder system. Courtesy of ILC Data Device Corp., Bohemia, N.Y.

Attaining high resolution is limited by mechanical as well as electronic constraints. However, if the space is available, a two-speed system can be used that will give high resolution and accuracy (Figure 5.16). In the two-speed system, there is a fine and a coarse position-to-digital conversion system such that the accuracy and resolution are improved by a factor of the gear ratio. For example, if a 14-bit (16,384-count) system with an 8-min accuracy is used as part of a two-speed system with a gear ratio of 32, then the resultant resolution becomes 19 bits, with an accuracy of 15 seconds.

The increase in cost generally does not justify acquiring superhigh resolution except in rare situations. Also necessary for consideration is the maximum rate at which the transducer shaft can turn and still allow the electronics to work effectively. The relationship between lead-screw pitch and linear-position displacement rate is related to the rotary tracking rate. If the linear displacement rate is kept constant while the lead-screw pitch is reduced (to gain higher resolution), then the tracking rate of the converter must increase. For example, if a converter has a maximum tracking rate of 12,000 rpm, then a lead screw with a 0.2-in. pitch can be displaced at a rate of 40 in./s, or 200 ft/min; however, if the pitch is changed to 0.1 in., then the displacement rate changes to 20 in./s, or 100 ft/min. When a resolver to digital converter is exposed to a slewing rate that exceeds the maximum tracking rate, it will lose tracking; but when the rate returns to the allowable limit, the converter will resume its normal operation.

The advantages and disadvantages of both optical encoders and resolver systems are summarized in Table 5.1.

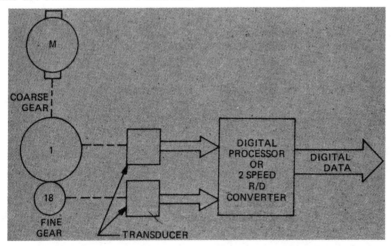

FIGURE 5.16. Accuracy and resolution can be improved with use of a two-speed system; if a gear ratio was 18 : 1 and the resolution of the coarse encoder system was 1.3 min, then the above two-speed system will increase the resolution to 0.075 min [1.3 min/18 = 0.072]. The accuracy will improve proportionally. Courtesy of ILC Data Device Corp., Bohemia, N.Y.

TABLE 5.1
COMPARISON OF OPTICAL AND RESOLVER ENCODERS

	INCREMENTAL OPTICAL ENCODER	ABSOLUTE OPTICAL ENCODER	RESOLVER SYSTEM
No. of Power Source Req'd.	1	1	2 or 3
Weight	Fair	Fair	Good
Size	Fair	Fair	Good
Vibration Resistance	Fair	Fair	Good
Temp. Effect	Fair	Fair	Good
No. of Output Lines	Low	High	Low
Mechanical Precision	Medium	High	Low
Absolute Position	No	Yes	Yes
No. of Basic Components	1	1	3
Velocity Output	No	No	Yes
Accuracy per Res.	Better	Better	Good
Cost	Low	High	Medium

Source: ILC Data Device Corp., Bohemia, N.Y. Used with permission.

When comparing optical encoders with resolver to digital systems, two separate comparisons must be made, because the optical encoder encompasses both mechanical and electrical components in one housing, whereas the mechanical (resolver) and the electronic component (converter) are separate in a resolver system. The fact that the optical encoder is a self-contained system is both its strength and weakness. The strength lies in the fact that all necessary elements for position sensing, digitization, and transmission of digital data are in one housing.

The resolver system requires a separate position sensor, electronic converter, and an excitation source. However, the optical encoder electronics must face the same environmental conditions as its mechanical elements, whereas the electronics of a resolver system can be kept in an area away from heat, vibration, and contamination. Also, because of its self-contained nature, the optical encoders are larger and cannot be located in a space-limited area such as a robot arm. In addition, the precision position-sensing elements of the optical encoder are less capable of withstanding harsh environmental conditions than an electromechanical resolver transducer.

Where excessive heat, vibration, or contaminants are not present, such as in coordinate-measuring machine applications, the optical encoder will operate well. The incremental encoder in the 10-bit (1024-count) range is especially advantageous because of its convenient self-contained nature and relatively low cost. However, if absolute position information is desired, the alternative is either an absolute optical encoder or a resolver system; the absolute encoder price is much higher than its incremental counterpart.

When the cost of an absolute encoder and a resolver system is considered, the encoder cost increases at a greater rate than a resolver system (Figure 5.17). The accuracy of the encoder system is generally $\pm \frac{1}{2}$ to ± 1 LSB, while the accuracy of the resolver system is determined by the resolver accuracy and the converter accuracy.

For a 10-bit resolver system, typical accuracy is about 1 LSB, while in a 14-bit system the accuracy is 6 to 7 LSBs or approximately 9 min. These accuracy considerations are exclusive of errors due to mechanical coupling of the position sensor to the primary shaft, which is a function of user requirements.

Recently, position control boards that can process various transducer outputs have become available. One manufacturer of machine tool controls has recently introduced a line of PC board assemblies designed to interface with different transducers. Unfortunately, the boards were designed only to interface with their own control electronics.

DDC's 5525 (Figure 5.18) is a three-axis control board that can interface with resolvers or Inductosyns, and comes with its own reference oscillator, multiturn counter, status register, D/As, and gain control for slewing. The control board includes all the electronics necessary for interfacing with the DEC-Q-Bus. Although there are control boards that can interface optical encoders

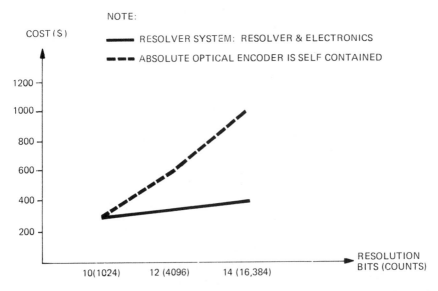

FIGURE 5.17. The sharp difference in the cost curve of the absolute optical encoder and the resolver/digital–encoder system is due to the fact that the 10-, 12-, and 14-bit resolver system can utilize the same basic encoder while the optical encoder system requires an increase in costly mechanical precision. The cost increase of the electronics for the resolver encoder is about $50 for every two additional bits. Courtesy of ILC Data Device Corp., Bohemia, N.Y.

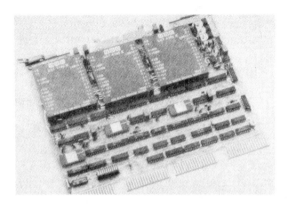

FIGURE 5.18. DDC–5525 Q-bus-compatible axis-control board. Axis control requirements for industrial position control generally appear in sets of three axes, an *X*, *Y*, and *Z* axis. The control board works as a three-channel resolver encoder system, and includes D/As and gain control for coarse and fine position control. Courtesy of the ILC Data Device Corp., Bohemia, N.Y.

with a popular microprocessor bus, the advantage of the resolver system on such a board comes from the fact that the rugged resolver can be placed in a harsh or space-limited environment, while the electronic control board can be placed in a protected environment.

In both the optical encoder and the resolver system, shielding and grounding precautions must be taken in order to maintain proper output position information. There are many different applications and many different requirements for shielding. Some users have designed special shielding for signal lines; others have run them in a trough designed to shield against electromagnetic noise; still others merely use twisted, shielded signal lines.

There is no single rule for protecting against all conditions; however, use of twisted shielded signal lines, grounded at one end only, and keeping the signal lines away from sources of electrical noise (welding machines, solid-state switches, change in inductive loads, high-current conductors) are good rules.

When a position sensor must be mounted in an area with high vibration, the resolver is a better choice because there are no precision coding disks, light sources, and electronics to be subjected to the vibration. In high-vibration environments the proper coupling of the transducer to the rotating shaft is important; set screws are generally not recommended.

Areas with heavy contamination (such as oil and dust) are a problem to optical encoders; special enclosures are available. Resolvers do not require special enclosures because inductive coupling is not affected by oil and dust.

The proper selection of position encoders requires consideration of resolution, accuracy, the mechanical and electrical conditions to which the transducer is to be exposed, ease of implementation, and cost.

The HSDC–8915 is a 14-bit hybrid microelectronic resolver or synchro to digital conveter. This position-to-digital converter is built around a custom large-scale-integrated (LSI) chip. The low component count within the hybrid accounts for high reliability in a rugged industrial environment. Available accuracy grades are 2.6 or 5.3 min with a repeatability of 1 LSB. The HSDC–8915 is packaged in a hermetically sealed 36-pin double dual-in-line package.

The SDC–19000 Series discrete resolver or synchro-to-digital converter is also available from the ILC Data Device Corp. in 10-, 12-, or 14-bit resolutions. Accuracy grades are 2.6, 8.5, 5.3, or 21 min. Tracking rates of up to 12,000 rpm are available for the 10-bit unit. The converters are potted for rugged industrial applications and costs start at $149 (in 100's).

5.11   Generalized Data Acquisition and Conversion Systems[h]

Figure 5.19 shows the digital measurement block diagram of a sensor, the parameters required to convet one variable to another, and the mechanism required to obtain the digital signal. Data acquisition and conversion systems include all the electronic devices from the sensor output requiring interfacing to the readout instruments used for visualizing the electronic changes. The conversion process converts the analog signal to a digital one and reprocesses it to a digital or analog output signal.

In the most general sense, interfacing can be regarded as a process of matching together dissimilar forms of information. At one end of the scale, this can be something as rudimentary as simply transposing wires, but on the other hand it can be a quite complex piece of circuitry, for example, a multi-channel ADC (analog-to-digital converter). Some of the principal techniques that properly come under this head are the following:

1. Wiring transposition
2. Impedance matching
3. Voltage-to-current conversion
4. Current-to-voltage conversion
5. Level adjustment
6. Analog-to-digital conversion (A/D or ADC)
7. Digital-to-analog conversion (D/A or DAC)
8. Data buffering (speed matching)
9. Multiplexing (speed matching)
10. Latching
11. Isolation

---

[h] The material in Section 5.11 is used courtesy of Feedback, Inc., Berkeley Heights, N.J.

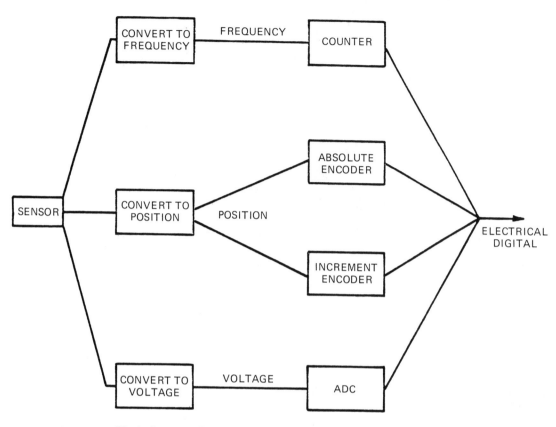

FIGURE 5.19. Block diagram of a digital measurement.

Figure 5.20 illustrates a selection of these techniques in symbolic form only. These are not necessarily practical operating systems.

All ADCs make use of a DAC within a feedback loop of some type. The actual techniques available are numerous, and the choice depends greatly on economic and suitability considerations. Figure 5.21 shows three of the available methods, and other general areas of application are as follows:

1. *Staircase type.* This is a relatively slow method and the conversion time is variable, but a development of it known as a *tracking converter* is ideal for the continuous conversion of slowly changing variables.
2. *Dual slope.* This is a relatively inexpensive technique but not very fast. It is widely used in digital voltmeters.
3. *Successive approximation.* This has the advantage of a constant conversion time regardless of the magnitude of the input voltage and is relatively fast; because of a resolution of one in two to the N (n inputs), the number of steps is only N instead of a possible two to the N.

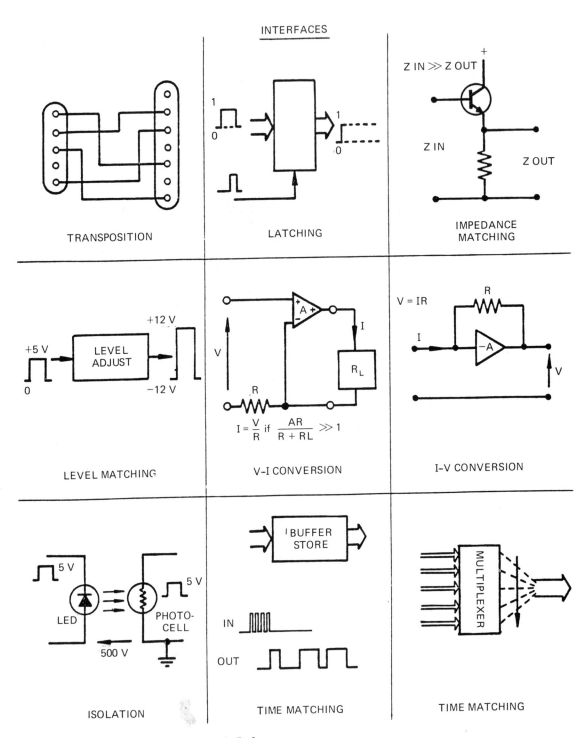

FIGURE 5.20. Interfaces in symbolic form.

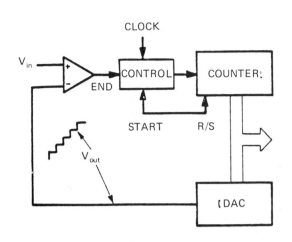

1. STAIRCASE/COMPARE

(ALSO USED IN TRACKING TYPE)

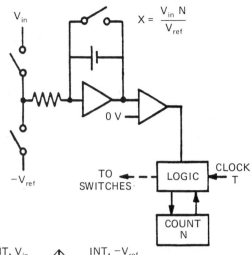

$$X = \frac{V_{in} N}{V_{ref}}$$

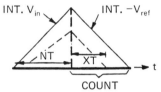

2. DUAL-SLOPE

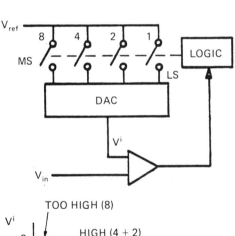

3. SUCCESSIVE APPROXIMATION

FIGURE 5.21. ADC methods.

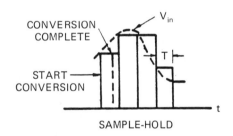

SAMPLE-HOLD

IF V_{in} CHANGES BY LESS THAN 1 LSB DURING CONVERSION TIME NO SAMPLE HOLD NEEDED

SAMPLING THEOREM

IF MAXIMUM FREQUENCY COMPONENT OF V_{in} IS fm THEN

$$T \text{ MUST BE} < \frac{1}{2fm}$$

Figure 5.22 also shows the meaning of sample and hold and the limits on sampling speed detected by the sampling theorem. If the latter are not observed, the phenomenom of *aliasing* occurs. The successive-approximation ADC method can, with quite moderate clock frequencies, cope with fairly rapid changes in input voltage without the necessity for a sample–hold circuit. For example, a 250-Hz input frequency would not be difficult to achieve.

Figure 5.23 is a simplified representation of a general-purpose ADC–DAC interface. The ADC employs the technique of *handshaking*, which is very common in many types of interface. The processor signals the ADC to commence, and the ADC in turn signals the processor when conversion is complete. The command port can be set up so as to choose either ADC or DAC at a given time and also to determine the direction of the data port.

Two of the most important interfacing techniques are ADC and DAC. These devices have become available in integrated solid-state packages that contain most, if not all, of the facilities needed in interfacing.

Figure 5.23 shows two basic principles of the DAC (digital-to-analog converter). They are not practical circuits as shown. For instance, what are shown as contacting switches will in fact be highly efficient, low offset transistors or metal oxide silicon field-effect transistor switches.

Method A in Figure 5.23, the weighted current principle, is not much used in practice nowadays because it demands a very wide range of resistor values, and the most significant values (lowest resistance) must be constructed to a very high accuracy if the overall conversion is to remain monotonic. Figure 5.23C shows the meaning of this term and other sources of error in digital-to-analog conversion.

The ladder network in Figure 5.23B is almost universally used, as only two values of resistor are needed and the demands on the switching transistors or other devices are more constant.

The DAC errors shown as linear in the diagram can be trimmed out by suitable adjustments in the circuit, but those classified as nonlinear cannot be. All errors can vary with temperature. Some typical figures for an 8-bit DAC are the following:

1. Linear error of ± 0.5 least significant bit (LSB) and temperature coefficient of ± 0.3 ppm/°C.
2. Differential nonlinearity of ± 0.5 least significant bit and temperature coefficient of ± 0.6 ppm/°C.
3. Offset voltage of 2 mV and temperature coefficient of ± 0.6 ppm/°C.
4. Full-scale output of 2.55 V.
5. Output impedance of 4000 Ω.
6. Settling time to 0.5 LSB of 0.8 μs.

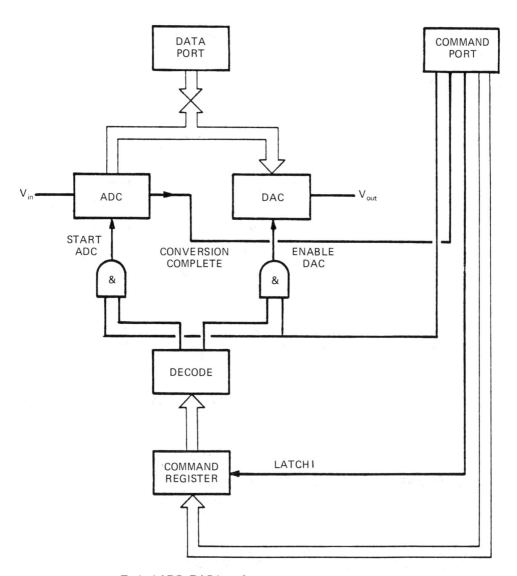

FIGURE 5.22. Typical ADC–DAC interface structure.

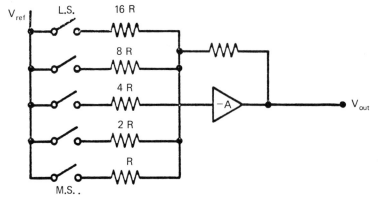

(A) WEIGHTED CURRENT DAC

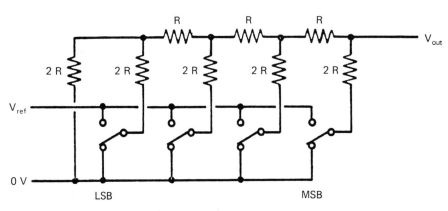

(B) LADDER NETWORK DAC

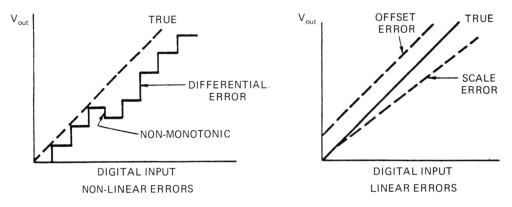

(C) DAC ERROR MECHANISMS

FIGURE 5.23. Basic principles of the DAC.

The binary inputs to DAC are generally taken through latches so that the microprocessor port does not need to hold the input and can be used for other purposes once the DAC input has been established.

5.12   Computer Bus Interfacing
for Industrial Robotics[i]

Many robots use the RS–232–C interface to or from a microcomputer in a robot system. This serial interface standard requires only one twisted line and sends out a bit at a time. Being serial, the user can drive lines a further distance with the RS–232–C. Modems also use a variation of the RS–232–C terminal. Where high noise environments and speed application are important considerations, the IEEE–488 should be used.

The RS–232–C standard developed by EIA Industries of Washington, D.C., is entitled the "Interface Between Data Terminal Equipment (DTE) and Data Communication Equipment Using Serial Binary Exchange." The RS–232–C standard was documented on March 9, 1979. The robotic and medical industries use the RS–232–C interface. The RS–449 standard is used in new equipment design for a general 37-position and 9-position interface for data terminal equipment and data-circuit terminating equipment using serial binary interchange. The binary symbol 0 or 1 is called a bit (binary digit).

In an industrial robotic system, two or more robots can be connected in series via the RS–232–C terminals and the robots operated from the same microcomputer or terminal.

IEEE–488 standard defines mechanical, electrical, and functional standards for achieving proper and efficient interfacing (interconnecting) of equipments via a bus. Important considerations include the following:

1. All data must be digital.
2. The number of devices that may be connected to one bus must not exceed 15.
3. The length of interconnecting cable (i.e., the transmission path length) must not exceed 20 meters (or two times number of instruments, in meters).
4. The rate at which data can be transmitted over the interface must not exceed 1 megabit/second (1 Mb/s = 10^6 bits/s).

The interface is byte serial/bit parallel; that is, each byte of 8 bits is transmitted serially, but all bits of each byte are transmitted simultaneously in parallel.

---

[i] The material in Section 5.12 is used courtesy of National Instruments, Austin, Texas.

The interface shown in Figure 5.24 provides high-speed operation for demanding applications. A multiple computer system allows flexibility because intrumentation can be shared by several processors.

Handshake is a proper sequence of status and control signals for transferring data via the interface using a data ready/data received technique. The three handshake lines provide a means of synchronously transferring data between instruments.

The IEEE P 728 standard establishes guidelines for the IEEE–488 code and format conventions for programmable measuring format.

5.13   Robotic Sensors Used in the National Bureau of Standards[j]

For robots to effectively operate in the partially unconstrained environment of manufacturing, they must be equipped with control systems that have sensory capabilities. Figure 5.25 shows a simplified block diagram of robot sensory systems with an afferent–efferent chain.

It is impossible to deal with control levels on one level because of the complexity of such a variety of sensory, control, and knowledge base issues. Only if the problem is partitioned into module components and the components ar-

FIGURE 5.24. PDP–11 minicomputer-to-GPIB interface; Model GPIB direct memory access [DMA]. Courtesy National Instruments, Austin, Texas.

---

[j] The material in Section 5.13 is used courtesy of the National Bureau of Standards, National Engineering Laboratory, Center for Manufacturing Engineering, Industrial Systems Division, Metrology Building, Room A127, Washington, D.C.

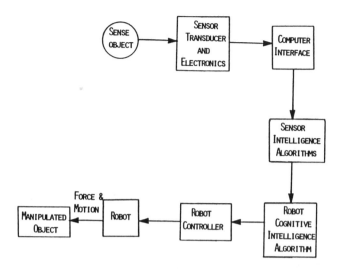

FIGURE 5.25. Block diagram of robot sensory systems with an afferent–efferent chain.

ranged in a hierarchical structure can the control issues be isolated into manageable units.

Figure 5.26 shows a microcomputer network designed at the National Bureau of Standards for a hierarchical control structure through a common bus.

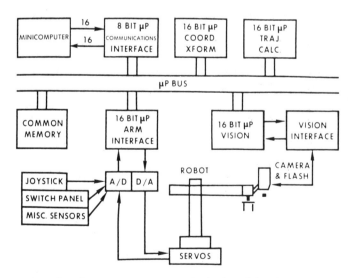

FIGURE 5.26. A microcomputer network designed at the National Bureau of Standards for a hierarchical control structure. Courtesy of the SPIE and the National Bureau of Standards.

The hierarchical control system developed at the National Bureau of Standards is a finite-state machine implemented on multiple microcomputers communicating through a common bus. Transformations of the coordinate system are shown in one of the microcomputers. The elemental move trajectories are calculated in a second microcomputer. The processing of the vision data is developed in a third microcomputer. The processing of force and touch data is placed in the fourth microcomputer (arm interface). A fifth microcomputer provides communication with a minicomputer wherein reside additional modules of the control hierarchy. A sixth microcomputer will be required for additional modules.

Communication from one module to another is accomplished through a sensory "mail drop" system. No two microcomputers communicate directly with each other. This means that common memory contains a location assigned to every element in the input and output sectors of every module in the hierarchy. In the future as many as ten or hundreds of microcomputers will be devised for control of the entire factors. In this manner, robots, machine tools, material transport systems, inventory control, safety and inspection systems can be integrated into a sensory-interactive, goal-seeking hierarchical computing struture for a totally automatic factory.

5.14   Robot and Operator Safety[k]

Robot safety objectives include the following:

1. Provide safe working conditions for the work place.
2. Ensure the integrity and safety of the robot to prevent damage to itself or objects that it manipulates.
3. Guarantee the safety of other machines in the robot environments.
4. Use proper robotic transducers to maintain proper safety requirements.

Robot safety requirements include the following:

1. Fail-safe behavior in all areas where robots are placed.
2. Safety-oriented configuration of the robot work envelope.
3. Allow robot intervention only by authorized personnel.

Fail-safe behavior requires emergency cutoff devices, including robotic transducers. The emergency cutoff device itself should not induce accidents and involves the following:

---

[k] The material in Section 5.14 is used courtesy of Feedback, Inc., Berkeley Heights, N.J.

1. Creep speed without drive energy.
2. Freeze motion may not always be the best response to an emergency cutoff.
3. Decision-making intelligence to produce an optimal response.

Power-failure modes include the following:

1. Adequate braking even during a power failure.
2. Avoid collapsible effect of unwanted movements when there is a power failure.
3. No automatic restart after a power failure or emergency cutoff.

5.14.1 Safety Light Curtain for Industrial Robotics[1]

The STI Model P4000 OPTOSAFE™ (Figure 5.27) is a programmable safety light curtain consisting of a modular transmitter, receiver, and control package. This system utilizes a patentable AGC (automatic gain control),

FIGURE 5.27. Miniaturized programmable light curtain. Courtesy of Scientific Technology, Inc., Mountain View, Calif.

---

[1] The material in Section 5.14.1 is used courtesy of the Scientific Technology, Inc., Mountain View, Calif.

which allows for the infrared transmitter and detector to compensate for changes in ambient atmospheric conditions such as direct sunlight, fog, dust, smoke, and other airborne particulates.

Tristate logic located within the P4000 allows optional control via a computer. The P4000 also contains special fail-safe and self-checking circuitry that monitors the system for component failures.

The Model P4000 Series OPTOSAFE has many applications. Many manufacturers are looking for ways to protect their employees from robots. This device is used to form an infrared curtain of light completely around the perimeter of the robot system, therefore making an invisible barrier for human safety. If any one of the beams is broken, the P4000 will signal the robot to stop. The P4000's individual channels are programmable, allowing the user to engage or disengage any channel of his selection. This option allows the user to program the robot to penetrate through a section of the curtain beam without triggering the robot to stop.

5.15   Review Questions

1. Discuss the rationale for force, torque, and motion transducers use in industrial robotics.
2. Discuss where proximity transducers are used in industrial robotics.
3. Discuss the merits of optical encoders for industrial robotics.
4. Discuss the merits of the resolver for industrial robotics.
5. Discuss the resolver to digital converter.
6. Given five machine functions in a factory, draw a block diagram showing a robot and where you place desired microcomputers to perform the machine operations properly. Do you need more than one robot?
7. Discuss potentiometric transducers for industrial robotics.
8. Discuss pulsed infrared photoelectric control in industrial robotics.
9. Discuss the use of the encoder interface module.
10. Compare optical and resolver encoders.
11. Discuss the data acquisition and conversion systems used for robotic transducers.
12. Discuss the methods used in analog-to-digital conversion.
13. Discuss the methods used in digital-to-analog conversion.
14. Discuss the RS–232–C and IEEE–488 bus interface.
15. Discuss the merit of the safety light curtain for industrial robotics.

5.16 Bibliography

John A. Allocca and Harold E. Levenson: *Electrical and Electronic Safety*, Reston Publishing Co., Reston, Va., 1982.

John A. Allocca and Allen Stuart, *Electronic Instrumentation*, Reston Publishing Co., Reston, Va., 1983.

John A. Allocca and Allen Stuart, *Transducers: Theory and Applications*, Reston Publishing Co., Reston, Va., 1984.

Geoffrey Boyes, *Synchro and Resolver Conversion*, Analog Devices, Norwood, Mass., 1980.

William Cullum and Harry Kratzer, "Measuring Shaft Position by Applying Synchro or Resolver Transducers," *Control Engineering*, Technical Publishing, © 1982.

Patrick T. Donahue, "What to Look For in a Manufacturer's IEEE–488 Specifications," *Measurements and Control*, Dec. 1982.

Steven Muth, "Consider Basic Parameters When Choosing S/D Converters," *Electronic Design News*, June 23, 1983, pp. 125–129.

The 1983 Industrial Sensor Directory™, Technical Database Corporation, Conroe, Texas, published yearly.

Stephan Ohr, *Data Converters for Robotics Application*, Electronic Products, New York, Sept. 7, 1983, pp 69–75.

Bart Sakasi, "Shaft Encoders Absolute Position Shaft Encoders for Control Systems," *Measurements and Controls*, Sept. 1982, pp. 178–182.

Synchro Conversion Handbook, 3rd printing, ILC Data Device Corporation, Bohemia, N.Y., 1982.

6

Mechanical Systems of Robots

6.1   Introduction

In this chapter, the mechanical features of industrial robots are highlighted. Grippers, harmonic drives, ball screws, and shock absorbers are discussed. The features of the mechanical systems illustrated should be used to visualize any problems encountered with the terminology used with industrial robots.

An interesting point in factory automation is the great number of mechanical parts that are handled by industrial robots. As an example, consider the automobile, with a large four-wheeled assembly weighing in excess of several thousand pounds; the greater number of these parts are processed as bits and pieces weighing less than several pounds. Countless small bits and pieces spend time waiting to be transferred to the next station.

Robots rely on fail-safe brakes to keep them in position and provide a safe operation. Fail-safe brakes stop a load whenever the coil is de-energized or turned off. Once the desired position is reached, power to the brake is turned off and the robot freezes. Then, when the robot has completed an operation and has to move away, power to the brakes is energized and associated servo-motors are used to produce the desired movement.

The components of the mechanical structure of robots include joints, links, end-effector (grippers), actuators (discussed in Chapter 4), and sensors (discussed in Chapter 5).

Robot arms generally include the following types:

1. Anthropomorphic arms that can bend and swivel at the "shoulder" and bend at the "elbow." These are the most flexible of robotic arms, capable of reaching nooks and crannies that others cannot. They are used in paint spraying.
2. Cylindrical arms that consist of two basic parts and include an extendable arm and the central pole on which it is mounted. The arm can move horizontally out from the pole, swivel around it, and move vertically up and down along it.
3. Polar arms consist of an extendable arm mounted on a central pivot. Like a cylindrical arm, a polar arm can swivel around its mounting. Instead of moving up and down it, however, the arm tilts to reach out above or below the level at which it is mounted. Spot welding of auto parts uses polar robot arms.
4. Cartesian arms use a gripper to move along three different, perpendicular tracks; one controls height, the second width, and the third depth of operation. This configuration gives the robot's arm great accuracy but makes it slow.

6.2   Objects, Position, and Time in Three Dimensions[a]

In spite of living and working in a three-dimensional world, it is often difficult to describe an object or define its position in three-dimensional space. Describing three-dimensional features on two-dimensional paper is even more difficult.

Technical people use a method called "perspective drawing" to describe an object in three dimensions. They also use a system called "coordinates" to define positions in three-dimensional space. Both methods are used here.

The letter Z identifies the vertical axis in a cartesian coordinate system, but the letter Y identifies the vertical axis in two-dimensional work. This apparent conflict is no problem to those accustomed to working in this area.

However, in the interest of simplicity and uniformity, the letter Y is used to identify the vertical axis in all cases, whether two or three dimensional.

The idea of *motion* contains two primitive concepts, those of *position* and *time*.

Strictly speaking, we should refer to position in some primary reference system that relates to a location in the universe. However, for practical rea-

---

[a] Sections 6.2 to 6.6 are from *Robotics . . . Start Simple, B.A.S.E.*, courtesy of the Mack Corp. Flagstaff, Ariz.

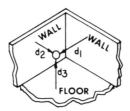

FIGURE 6.1. Position in a room. Courtesy of Mack Corp.

sons, we may do this by noting its distance from three mutually perpendicular planes, which we can take as a fixed reference system. An example would be the lines of intersection of two adjacent walls and the floor of a room. The position of an object in a room, then, is found by measuring its distance from the walls and floor (see Figure 6.1).

Any attempt to define what is meant by "time" or by "an interval of time" in terms of subjective phenomena leads to a maze of philosophic problems. In the real world, a measurement of time interval is more practical than knowledge of exact time. Multiples of time in terms of seconds, minutes, and hours are all that is required and can be achieved with precision from timing devices.

For practical considerations, motion can be defined as the process of changing place from position A at time 1 to position B at time 2. The path taken from path A to point B is also considered (see Figure 6.2).

Controlled-path motion follows a prescribed pattern and requires considerable sophistication in both the manipulator and the control system. An example is a robot following a curved member on a bicycle frame during a spray-painting operation.

Point-to-point motions, without regard to actual path, are much simpler. These applications involve only translation and rotation with simple se-

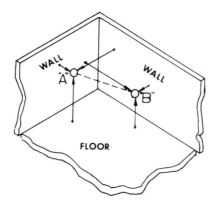

FIGURE 6.2. Motion shown from point A to point B. Courtesy of Mack Corp.

quence control. An example is a robot picking an object from a hopper, placing it at another location for processing, and then moving it to a conveyor.

Using robots with controlled-path capabilities in point-to-point transfer may be "overkill," while point-to-point robots may limit potential use for future applications. Appropriate economic study can make this judgment.

As to types of industrial robots, there are really only two categories: servo controlled and nonservo controlled. The nonservo robots are inexpensive, easy to understand and set up, have good precision with high reliability, and can be a logical choice for point-to-point transfer in manufacturing processes and assembly operations. Nonservo robots usually operate along the axes of a cartesian coordinate system with one or more rotations, such as wrist motion. Motion is controlled through the use of a limited number of stops, and the actual path between points may be difficult to define. The control system can be as simple as a cam or drum timer that sequences valves or switches.

Servo-controlled robots have a wide range of capabilities. They can perform multiple point-to-point transfers and move along a controlled path. Most servo-controlled robots use a jointed arm mechanism and can be programmed to avoid an obstruction. Programming can be rather sophisticated, and the price tag for a complete servo-controlled robot system is relatively high.

Regardless of type, an industrial robot must include the combination of manipulator and control system. The manipulator is the portion that goes through the motions and performs work. It can be as simple as a gripper on the end of a cylinder or so complex it takes a full-time staff of professional people to install and maintain a single system. The control system directs the manipulator in performing its work. A control system can be as simple as a clock-driven timer or so sophisticated that computer software to operate it is nonexistent at this time.

6.3   End-Effectors [Work-holding Devices]

End-effectors are devices that pick up, grasp, or otherwise handle objects for transfer or while being processed. Selecting an end-effector early in a program involving robotics is usually not difficult, since the characteristics of the object to be processed are well known.

End-effectors come in all types and sizes. Many must be custom made to match special handling requirements. Some typical examples are shown in Figure 6.3.

6.4   Grippers

Grippers are perhaps the most universal of all end-effectors. In many cases, reprogramming involves only a change in finger style (see Figure 6.4).

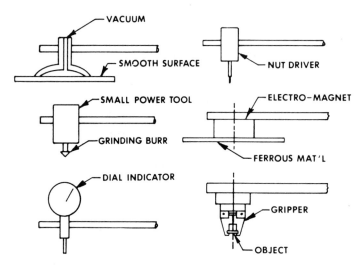

FIGURE 6.3. End-effectors. Courtesy of Mack Corp.

In the B.A.S.E.® robotics system, there are standard grippers in four sizes, in two- and three-finger configurations for external and internal gripping, as well as soft blank fingers that can be easily modified for special shapes (see Figure 6.5).

Advanced grippers use a three-fingered gripper. The three-fingered gripper results in added dexterity in gripping objects. Grippers need "intelligence" to determine the positioning of the "fingers" in the object.

Three-finger models duplicate the motions of the thumb, index finger, and a third finger for grasping bodies of revolution and objects of spherical or

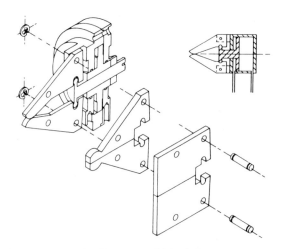

FIGURE 6.4. Gripper example. Courtesy of Mack Corp.

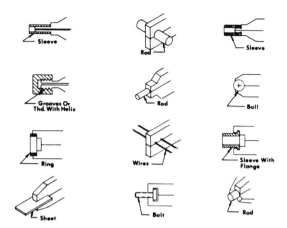

FIGURE 6.5. The B.A.S.E. robotic system showing grippers. Courtesy of Mack Corp.

cylindrical shape. Both versions have the self-centering feature of a standard two- and three-jaw chuck with repeatability within a few thousandths of an inch. Fingers are interchangeable in each of four model sizes and provide for rapid change-over when robots are reprogrammed for another task.

A pilot diameter and bolt circle are provided on the rear face for mounting. Simple adaptors will mount these grippers to almost any linear or rotary motion device to transport objects that are within the operating range of finger travel.

Transport can be as simple as mounting a gripper to the rod of an air cylinder for straight-line motion or mounting to a more complex device, such as to the "business end" of a six-component manipulator, to reach any point in a cartesian coordinate system.

All units operate on the principle of a double-acting cylinder controlled through a simple four-way valve circuit. Fluid pressure opens or closes the finger for positive operation. Maximum operating pressure is 150 psi in either hydraulic or pneumatic service; however, most applications use plant air at 80 psi for a simple and reliable source of fluid power.

Pinch force at 80 psi is in the range of 5 to 50 lb. depending on model size. However, pinch force is proportional to operating pressure and can be reduced to any nominally lower level for handling extremely delicate parts. Furthermore, pinch force is essentially constant for any given operating pressure, which allows the gripper to grasp objects of different size with uniform force.

An adjustment is provided to increase the dimension across the fingers in the open position, permitting larger objects to be handled when it is not necessary for the fingers to touch in the closed position.

Overall construction is as light as practical for high-speed cycling and rapid transport. The smallest units weigh 2 oz and the largest units weigh 9 oz. Pistons and adjusting stems are stainless steel; bodies, covers, and fingers are made from high-strength aluminum alloy. Seals are standard O-rings. Barbed fittings for 1/8-in. flexible tubing are included with each unit.

Having selected a gripper or other end-effector, it is time to consider mounting. Mechanical attachment is easy, but orientation requires more thought. To "start simple," think in terms of gripper orientation relative to the position of the part and the direction of approach and withdrawal.

As an example of orientation, in order to pick a ball from an open track, it is probably best to mount a gripper downward, open the fingers, transport downward, close the fingers, transport upward to clear the track, and proceed with subsequent translations and/or rotations (see Figure 6.6).

6.5   Transporters

B.A.S.E.® X Axis Transporters (Figure 6.7) are basically single-ended, double-acting air cylinders with special features for applications involving transfer of small loads over short distances (bench-top robotics).

Y and Z Axis Transporters are motion conversion mechanisms for transferring small loads over short distances along the Y and Z axis of a B.A.S.E. coordinate system.

Units are operated from plant air and many be controlled with simple air logic or with programmable controllers and computers when more sophistication is required.

Various motion possibilities can be assembled according to user requirements. Figure 6.8 shows a simple combination of a gripper with X and Y Axis Transporters for transferring an object in programmed motion in the XY plane.

Figure 6.9 shows a gripper in combination with X and Z Axis Transporters for typical motion patterns in the XZ plane of a B.A.S.E. coordinate system.

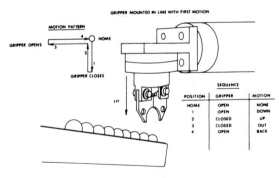

FIGURE 6.6. Mounting a gripper downward, open the finger, transport downward to clear the track, and proceed with subsequent translation and/or rotation. Courtesy of Mack Corp.

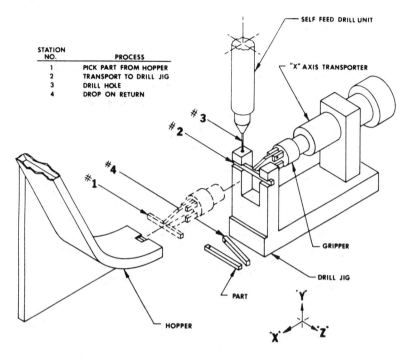

STATION NO.	PROCESS
1	PICK PART FROM HOPPER
2	TRANSPORT TO DRILL JIG
3	DRILL HOLE
4	DROP ON RETURN

FIGURE 6.7. X-axis transporter example used for a single-axis transfer. Courtesy of Mack Corp.

Figure 6.10 shows a combination of three B.A.S.E. transporters, a gripper, and an intermediate stop cylinder for programmed motion in the XYZ coordinate system. The motion patterns shown in this figure are typical of many that can be programmed with only a day or so of study by a novice who has access to a programmable controller. Although the robot shown in Figure 6.10 is very simple (no rotations), it can provide enormous benefits in better quality and increased productivity.

6.6   Rotators

Rotators are also key components in the B.A.S.E. Robot System. Rotators provide angular motion about the X, Y, and Z axis of a B.A.S.E. coordinate system.

The combination of three translations and three rotations comprises a full six-component robot system and provides all the motions necessary to transport something in space. In the B.A.S.E. system, rotators are identified as *roll*, *pitch*, and *yaw* and simulate the motions of the hand at the wrist (see Figure 6.11).

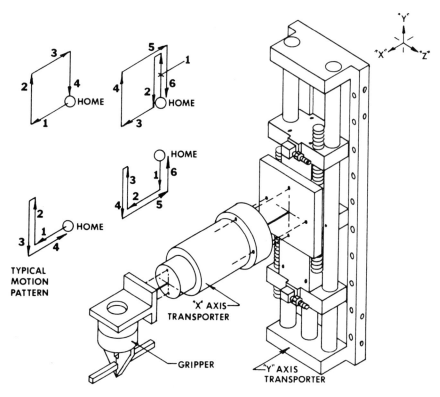

FIGURE 6.8. A simple combination of a gripper with an *X* and *Y* axis transporter for transferring an object in programmed motion in the *XY* plane. Courtesy of Mack Corp.

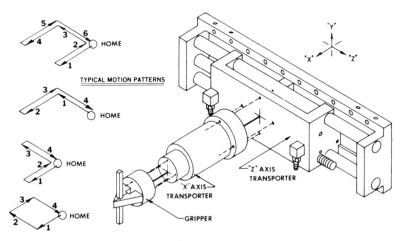

FIGURE 6.9. Gripper in combination with *X* and *Y* axis transporters for typical motion patterns. Courtesy of Mack Corp.

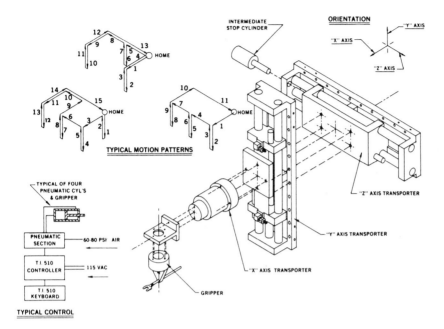

FIGURE 6.10. B.A.S.E. *X, Y,* and *Z* transporter shown with intermittent stop cylinder for programmed motion in the *XYZ* coordinate system. Courtesy of Mack Corp.

Rotators providing roll, pitch and yaw are installed between a *transporter* and *gripper* when needed. The rotators shown in Figure 6.12 duplicate the motions of Figure 6.11 mechanical form for simple robot applications.

Series 2, B.A.S.E. rotators are nonservo, air-operated, vane-actuated units with a choice of 90° or 180° rotations in roll, pitch, and yaw. Adjustable stops provide close control over angular position; however, stopping at some intermediate angle (such as $22\frac{1}{2}°$) is not practical without servo feedback or remotely controlled stops.

6.7 Mechanical Construction
of the Microbot Alpha Industrial Robot[b]

The essential elements of the Microbot Alpha robot (Figure 6.13) are as follows:

1. Motor drive circuit board housed in the base of the robot
2. Cast aluminum base

---

[b] Sections 6.7 to 6.9 are from *Microbot Alpha Volume 1 Reference Guide* (preliminary), by John W. Hill, Ph.D., of Microbot, Inc., © 1982 and is used courtesy of Microbot, Inc., Mountain View, Calif.

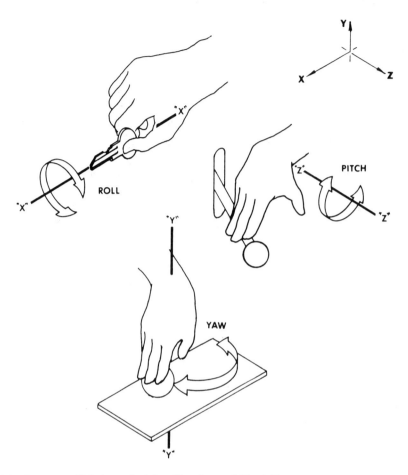

FIGURE 6.11. Rotator principles. Courtesy of Mack Corp.

3. Body
4. Upper arm
5. Forearm
6. Wrist joint
7. Hand (or gripper)
8. Six stepper motors
9. Timing belt/cable drive cluster

Protruding from the rear of the base is the robot cable, which plugs into the controller.

The body swivels relative to the base on a hollow shaft that is attached to the base. This is called the base joint. Six stepper motors with their corresponding timing belt/cable drive capstans are mounted on the body and con-

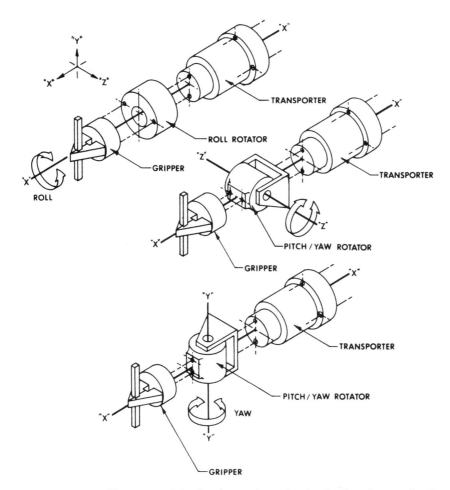

FIGURE 6.12. Demonstration of rotators in action in simple robot applications. Courtesy of Mack Corp.

trol each of the joints. The power wires, homing sensor wires, and pneumatic tubing pass from the interior of the base, through the hollow shaft, and into the body.

At the upper end of the body is attached the *upper arm*. The upper arm rotates relative to the body on a shaft; this shaft is called the *shoulder joint*. Similarly, the *forearm* is attached to the upper arm by another shaft known as the *elbow joint*. Finally, the *hand* (also called the end-effector or gripper) is attached to the forearm at the *wrist joint*. There are two motors for the wrist; one operates the left wrist joint, and the other operates the right wrist joint. Together, these two joints control the pitch and roll of the hand.

The Microbot Alpha has a lifting capacity of 1.5 lb when fully extended, and a worst-case resolution (the smallest amount the arm can be made to

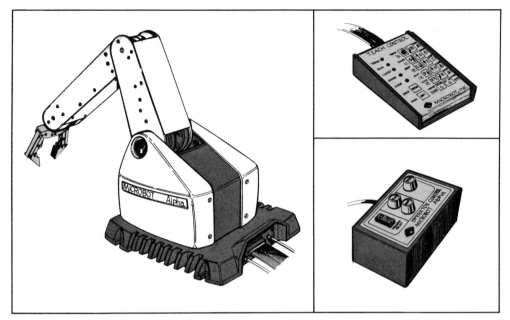

The MICROBOT Alpha is a low-cost, programmable, robot designed to operate continuously in a normal manufacturing or production environment. Its ruggedized construction permits handling of loads up to $1^1/_2$ lbs. at speeds up to 20 ips.

The MICROBOT Alpha is an extension of the same design principles utilized in our proven developmental robot, the MICROBOT TeachMover. The use of timing belts coupled with metal pulleys in the Alpha transmit the torque of the stepper motors mounted in the body to the stainless steel cables that control the arm motion. This precision cable technology has proven to be a strong, reliable method to control arm motion without adding unwanted weight to the arm, thus producing a low-cost robot without sacrificing any of the features found in higher-priced robots.

The MICROBOT Alpha is a completely self-contained system consisting of the 5-axis arm mounted on a base casting which houses the motor driver circuitry, a separate control enclosure with the power supply and microprocessor circuitry, a removable teach control to simplify complex programming tasks and an operator control.

The MICROBOT Alpha offers the user two programming options. Using the teach control, the microprocessor can be programmed to retain 227 points in permanent memory or it can be programmed in conjunction with most computers through their RS–232–C Serial Interface.

FIGURE 6.13. Microbot Alpha robot. Courtesy of Microbot, Inc.

move) of 0.010 in. The end of the hand can be positioned anywhere within a partial sphere that has a radius of 18 in., as shown in Figure 6.14. The range of speed is from 0 to 20 in./s. The 1.5-lb lifting capacity is based on a 10,000-hour life. The robot is capable of lifting greater loads, depending, of course, on speed and position of the arm. Operation with greater than rated loads will reduce life expectancy. Detailed performance characteristics of the Alpha are given in Table 6.1.

In general, the base, the body, and all the extension members are hollow sheet-metal parts, which are light in weight but stiff. All members are con-

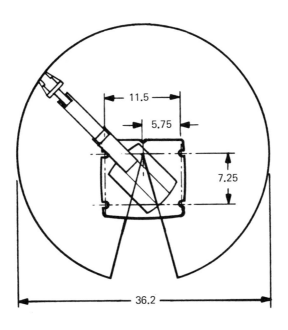

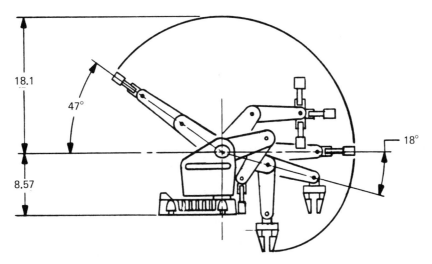

FIGURE 6.14. Operating envelope (measurements shown in inches). Courtesy of Microbot, Inc.

TABLE 6.1
MICROBOT ALPHA SPECIFICATIONS

System Description	Five axis robotic arm with cable operated mechanical gripper, and power supply/control interface.
Drive	Open loop control with automatic homing. Electrical stepper motors—half step pulses.
Controller	Microprocessor with 4K bytes each of EPROM and EEPROM and 1K byte of RAM.
Control Method	Coordinated point-to-point. Provision for segmented path control.
Programming Capacity	227-point permanent memory.
Computer Interface	Dual RS-232C asynchronous serial communications interface.
External Control Interface	Sixteen channels for optically isolated relays at factory voltage levels (8 Input, 8 Output).
Program Language	ARMBASIC® through serial port.
Power Requirements	110 VAC or 220 VAC.
Operating Environment	Ambient temperature limit 100°F (38°C). Internal mechanical working surfaces must be protected from debris or aggressive solvents.
Cable Lengths (to control panel)	Robot, 10 ft. (3m) Teach Control, 10 ft. (3m) Operator Control, 10 ft. (3m).
Payload	1.5 lbs. (681 gm).
Reach	18 in. (457 mm).
Speed	0–20 ips (0–508 mm/sec) selectable, with acceleration and deceleration.
Resolution	.010 in. (.25 mm) each axis.
Positioning Accuracy	±.020 in. (.50 mm).
Operating Envelope	Base 330° Shoulder 140° Elbow 140° Wrist Roll 360° Wrist Pitch 180°
Gripper	Standard cable operated mechanical gripper included with system. Force programmable 1 lb. (4.33 N) to 7 lbs. (86.7 N) Maximum jaw opening 3 in. (76.2 mm) See optional accessories for other grippers.
Optional Accessories	1) Teach Control, 14-key, 4-function, for on-site programming tasks. 2) Operator Control—for local program select, stop and reset by operator. 3) Pneumatic Grippers—internal and external, two finger and three finger plus blank fingers for custom applications. 4) Two extra stepping motors with drivers and control for user system integration of conveyors, rotating tables, part positioners, etc.

Source: Microbot, Inc. Used with permission.

nected to each other by means of shafts, or axles, which pass through bushings mounted on the members.

The Microbot Alpha is a low-cost, programmable, robot designed to operate continuously in a normal manufacturing or production environment. Its ruggedized construction permits handling of loads up to $1\frac{1}{2}$ lb at speeds up to 20 in./s.

The Microbot Alpha is an extension of the same design principles utilized in our proven developmental robot, the Microbot TeachMover. The use of timing belts coupled with metal pulleys in the Alpha transmit the torque of the stepper motors mounted in the body to the stainless steel cables that control the arm motion. This precision cable technology has proved to be a strong, reliable method to control arm motion without adding unwanted weight to the arm, thus producing a low-cost robot without sacrificing any of the features found in higher-priced robots.

The Microbot Alpha is a completely self-contained system consisting of the five-axis arm mounted on a base casting that houses the motor driver circuitry, a separate control enclosure with the power supply and microprocessor circuitry, a removable teach control to simplify complex programming tasks, and an operator control.

The Microbot Alpha offers the user two programming options. Using the teach control, the microprocessor can be programmed to retain 227 points in permanent memory, or it can be programmed in conjunction with most computers through their RS–232–C serial interface.

6.8 Cable Drive System of the Microbot Alpha Robot

The cable drive system of the Microbot Alpha employed to manipulate the arm members is unique in industrial manipulators. From the drive cluster in the body, aircraft-quality cables extend to the base, upper arm, forearm, and hand. This is shown in Figure 6.15. This cable design is an adaptation and refinement of the "tendon technology" used in aircraft, high-speed printers, and hot cell manipulators. Note that the cables are wound around the hubs of the drive pulleys. This serves not only to provide take up drums for the cables, but also gives the proper reduction ratios for all six drives.

All six drive motors have been mounted on the body. This allows the weight of the extension members to be kept at a minimum and reduces motor torque requirements. To reduce the number of moving parts, all six timing belt drives were mounted on the same shaft.

Each cable of the drive system has certain features in common. Only the routing of the individual cables is different. Each cable has a fixed attachment point. Next, a small ball is swaged onto the cable. This ball is designed to be press-fit into the drive capstan, providing positive drive capability and ease of

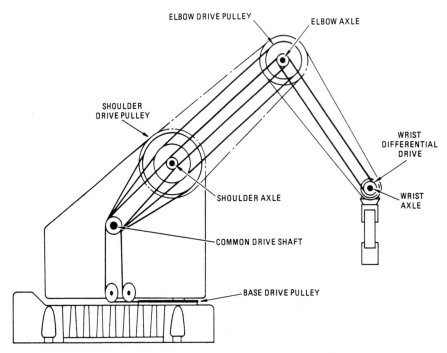

FIGURE 6.15. Simplified cabling diagram. Courtesy of Microbot, Inc..

assembly and/or replacement. Finally, the cable is terminated by a tension adjustment mechanism. Both ends of the cable are attached to the member to be driven. Anytime a cable passes over any pulley, that pulley is spirally grooved. This increases cable life and ensures that the cables remain on the pulley. The wrist cables are exceptions in that the fixed end of the cable is continued around and is attached to the adjustable end, which is not attached to the member.

1. *Base drive.* The body cable causes the body to rotate on the vertical base pivot by driving a large pulley welded to the base. Two small pulleys, located at the bottom of the body, change the base cable direction so that the cable feed is tangent to the surface of both the drive drum and the base pulley. The home sensor, located at the front of the base, is aligned to signal the center of the body rotation range.

2. *Shoulder drive.* The shoulder cable causes the upper arm to rotate relative to the body on the shoulder axle. Both ends of the cable are terminated on the upper arm. The cable drive system has the additional features that rotation of the shoulder joint causes equal and opposite elbow and wrist rotation so that the orientation of the hand remains unchanged. The home sensor for this joint is located in the shoulder.

3. *Elbow drive.* The elbow cable causes the forearm to rotate relative to the upper arm on the elbow joint. This cable first passes around an idler pulley on the shoulder axle, and then around a drive pulley of the same diameter, which is attached to the elbow axle. The cable then terminates on the forearm. Notice that, again, rotation of the elbow causes opposite and equal rotation of the wrist, thus maintaining hand orientation. Finally, note that manually moving the elbow joint causes the hand to open and close. This interaction is automatically decoupled in the Alpha software. The elbow home sensor is located at the elbow.

4. *Wrist drive.* The right and left wrist cables cause the hand to "roll" and "pitch" relative to the forearm. These cables together control the wrist joint (Figure 6.16). Both cables pass around idler pulleys on the shoulder axle and the elbow axle, and then around the hubs of miter gears located on the wrist axle. Tension is maintained in both cables by means of turnbuckles. Note that the two miter gears on the wrist axle mesh with the miter gear on the hand axle. This configuration forms a

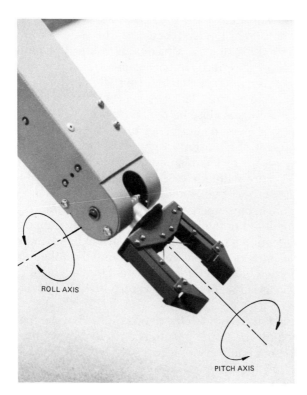

ROLL AXIS

PITCH AXIS

FIGURE 6.16. Roll and pitch axes. Courtesy of Microbot, Inc.

differential gear set. By driving the left and right wrist cables in the same direction, pitch is achieved; by driving them in the opposite direction, roll is achieved. Home sensors for the wrist drives are located inside the forearm.

6.9   Cable Gripper of the Microbot Alpha Robot

The operation of the Microbot Alpha robot is now described. The hand and hand cabling are shown in Figure 6.17. Attached to the output gear of the differential gear set, the hand housing holds two pairs of links, and each pair of links terminates in a *grip (finger)*. The housing, the links, and the fingers are attached to each other by small pins. Torsion springs located on the pins attach the links to the hand housing and provide the return force to open the hand as the hand cable is slackened.

The *grip cable* is attached to the hand drive drum located in the body. The cable passes over an idler pulley located on the shoulder axle. Tension is kept on this cable by means of a tensioner device located on the shoulder axle. From the shoulder idler, the hand cable then passes over an idler pulley mounted on a sensing bracket located in the upper inside of the upper arm. Mounted on the sensing bracket is a microswitch (or grip switch), which is activated by an increase in grip cable tension. This tension is built up when the hand closes on an object (or the fingers close on one another).

Finally, the hand cable passes over an idler pulley on the elbow shaft and one in the forearm and terminates at a tension spring. Attached to the other end of the tension spring, by means of a cable clamping device and in line with the hand cable, are two separate link-drive (finger) cables. These cables pass over two guide pulleys in the wrist yoke and then through the center of the hollow hand axle. When the cables emerge from the hand axle, they pass over separate idler pulleys mounted in the base of the hand. They then pass around idler pulleys that are mounted on the inner links of the hand and return to and terminate on the shafts of the two pulleys mounted on the hand base. This arrangement forms two block-and-tackle devices that augment the gripping force of the hand. The use of identical cabling on both links provides for symmetrical hand closure.

The tension spring mounted in series with the hand-cable drive assembly permits gripping force to be built up by the position-controlled drive motor once the hand has closed. The maximum gripping force is 7 lb; this occurs approximately 160 motor steps beyond the hand closure.

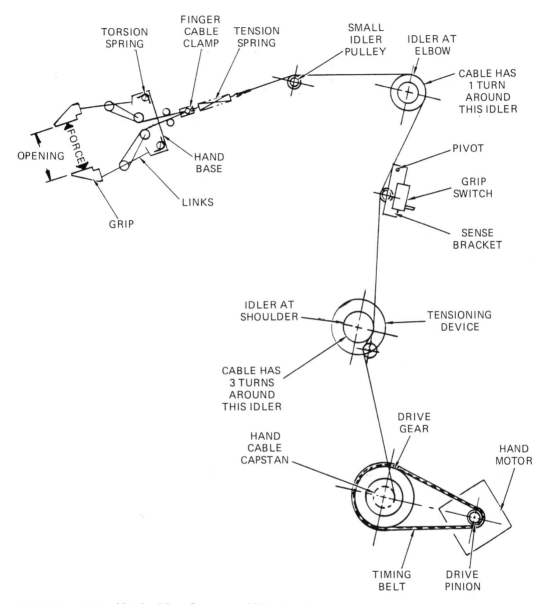

TORSION SPRING
FINGER CABLE CLAMP
TENSION SPRING
SMALL IDLER PULLEY
IDLER AT ELBOW
CABLE HAS 1 TURN AROUND THIS IDLER
OPENING
FORCE
HAND BASE
PIVOT
GRIP SWITCH
SENSE BRACKET
LINKS
GRIP
IDLER AT SHOULDER
TENSIONING DEVICE
CABLE HAS 3 TURNS AROUND THIS IDLER
DRIVE GEAR
HAND CABLE CAPSTAN
HAND MOTOR
TIMING BELT
DRIVE PINION

FIGURE 6.17. Hand cabling. Courtesy of Microbot, Inc.

171

6.10   Mechanical System of the IRI M50[c]

Each chain drive system of the IRI M50 of International/Robomation Intelligence for the torso, shoulder, and the elbow axis includes six stages: motor stage, four stages of sprocket ratio increase, and the joint stage. The chain drive system in the wrist module (forearm) contains the separate stages for the pitch and roll axis, each one having five stages: motor stage, three stages of sprocket ratio increase, and a joint stage. The chain sprocket subassemblies include sprockets, shafts, and end supports that house ball bearings. The subassemblies are retained by two hex head screws in each bearing housing. One screw on each side of the module is in a slotted hole of the module structure to permit chain adjustment. Chain adjustment is done by adjusting screws that contact the bearing housing one on each side. Different-sized chains of various lengths are used in each module. Chains utilize spring clip connecting links or offset links for installation.

The module description of the mechanical system of the IRI M50 is shown in Figure 6.18. The torso module is bolted at its forward end to the spindle box assembly. Its forward sides are bolted on one side to the equipment module and on the opposite side to the shoulder module. The torso module contains the chain stages for rotating the robot about its vertical axis. Access holes on the sides, tip, and bottom of the torso module provide access for inspection and maintenance such as chain adjustments. The air motor/valve assembly is mounted on top of the module at the aft end. The dual brakes are on the top and bottom of the sprocket at stage 1, while the encoder is installed at the motor stage.

The shoulder module is on the right side of the spindle box assembly and the forward right side of the torso module. It contains the chain stages for rotating the shoulder shaft subassembly. The right side of the shoulder shaft assembly is supported in the module by thrust washers and two thrust bearings and two radial needle bearings. Access holes are provided for chain adjustment and maintenance. The dual brakes are installed at stage 1, while the encoder is at the motor stage.

The elbow module hinges on the shoulder shaft and is supported by bathtub fittings bolted to the shoulder shaft subassembly. Side braces rotate on the ends of the shoulder shaft and are attached to the sides of the module for additional support. The module contains the chain stages for rotating the elbow shaft inside the two thrust bearings and two radial bearings installed in the forward end of the module. Access holes are provided for chain adjustment and maintenance. The dual brakes are installed at stage 1, and the encoder is installed at the motor stage.

---

[c]Section 6.10 is from *IRI M50 Robot Document Reference*, Courtesy of International Robomation/Intelligence, Carlsbad, Calif., © Jan. 27, 1983.

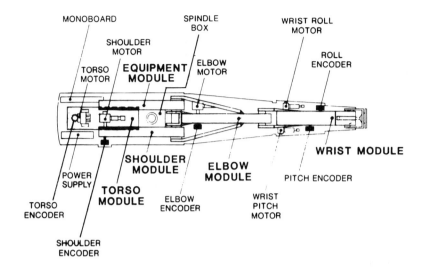

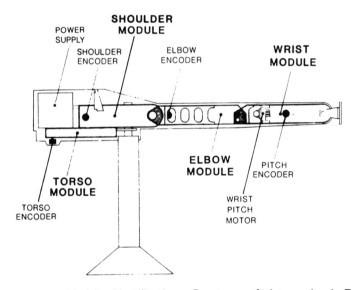

FIGURE 6.18. Module identification. Courtesy of International Robomation/ Intelligence, Carlsbad, Calif.

The wrist module is attached to the elbow shaft by means of a bolted clamp. It contains the chain stages for pitch and roll axes. Access to the chains for adjustment and maintenance is through access holes on the sides and top of the module. The air motor/valve assembly and encoder for the pitch are on the right side, while for the roll they are on the lift. Brakes are installed at stage 1 for each axis, and the encoders are at stage 2.

The chain drive for the pitch axis rotates the wrist shaft, which is installed inside the needle bearings and thrust bearings. The chain drive for the roll axis rotates the hand by a bevel gear arrangement. Access to the wrist shaft assembly is by removing the front cover on the module. A nylon filter prevents nominal foreign objects from entering the wrist shaft assembly.

The equipment module is installed on the left side of the spindle box assembly and the forward left side of the torso module. It contains the pressure control assembly and two needle bearing supports for the left side of the shoulder shaft. Access to the pressure control assembly components is by a removable top cover. This module also contains the monoboard assembly.

The spindle box assembly is bolted to the equipment, shoulder, and torso modules. It rotates around the spindle shaft, which is secured to the robot mounting base. A needle bearing and a thrust bearing on the spindle at the top and bottom of the box assembly carry the torso axis loads of the IRI M50. The thrust bearing on the bottom of the spindle rests on the torso joint sprocket, which is welded to the spindle shaft. The spindle shaft is retained in the box by a lock washer and nut that is tightened against the top thrust bearing.

6.11 ´   Ball Bearing Screws and Splines
Used in Robot Systems[d]

A ball bearing screw (Figure 6.19) is well-described by its name: it is a screw that runs on bearing balls. The screw thread is actually a hardened ball race. The nut consists of a series of bearing balls circulating in a similar race and carried from one end of the nut to the other by return tubes. The balls provide the only physical contact between screw and nut, replacing the sliding friction of the conventional acme thread screw with a free, rolling motion.

The primary function of a ball bearing screw is to convert rotary motion to linear motion—or torque to thrust. Linear motion may also be converted to rotary motion—or thrust to torque. This conversion is normally accomplished in two ways (see Figures 6.20 and 6.21).

Warner Electric Brake and Clutch Co. has products that can be utilized for motion control functions both on a robot and on equipment that operates in conjunction with the robot. This includes the following:

1. *On robot.* Component integral to a robot drive system or sensing system, for example, a ball bearing screw installed in robot manipulator arm to provide high efficiency extension, or fiberoptic photoelectric installed in gripper to sense presence of part.

---

[d] The material in Sections 6.11 to 6.13 is used courtesy of Warner Electric Brake and Clutch Co., South Beloit, Ill.

FIGURE 6.19. Ball bearing screw. Courtesy of Warner Electric Brake and Clutch Co., South Beloit, Ill.

2. *Off robot.* Component utilized in a piece of equipment that is dedicated to the robot operation, but not part of the robot, for example, a ball bearing screw in an XY positioning table (see Figure 6.22).

Ball bearing screws and splines are commonly used in the drive systems of electric robots to effect high-efficiency linear movement of the robot manipulator arm. The linkages and additional drive components used for each robot vary according to the particulars of each design. Ball bearing screws and

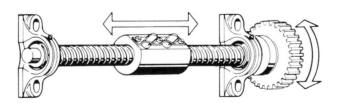

FIGURE 6.20. In the most common application, the screw rotates and the nut travels along the length of the screw. Courtesy of Warner Electric Brake and Clutch Co., South Beloit, Ill.

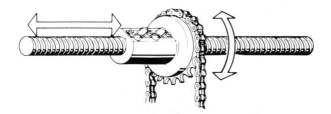

FIGURE 6.21. In some applications, the nut is rotated and the screw travels through the nut. Courtesy of Warner Electric Brake and Clutch Co., South Beloit, Ill.

splines can also be used in carriage devices to move the entire robot (see Figure 6.23).

Ball bearing screws and splines are also used in hydraulic and pneumatic robots. Although the predominance of cylinder actuators has limited their application somewhat in fluid power robot drives, several manufacturers have applied ball bearing screws and splines on their hydraulic machines.

Ball bearing screws are also used in the drive systems of jointed-arm robots to effect the elevation and reach movements of the manipulator arm (see Figure 6.24).

6.12   Electrically Released Brakes for Robots

Electrically released (ER) brakes, shown in Figure 6.25, are commonly applied on robots for the following reasons:

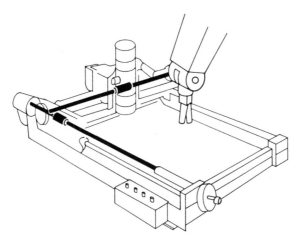

FIGURE 6.22. Ball bearing screw in an *XY* positioning table. Courtesy of Warner Electric Brake and Clutch Co., South Beloit, Ill.

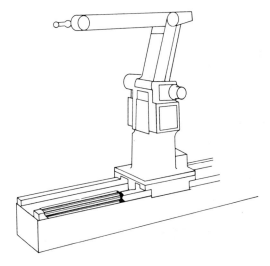

FIGURE 6.23. Ball bearing screws can be used in carriage drives to move the entire robot. Coutesy of Warner Clutch and Brake Co., South Beloit, Ill.

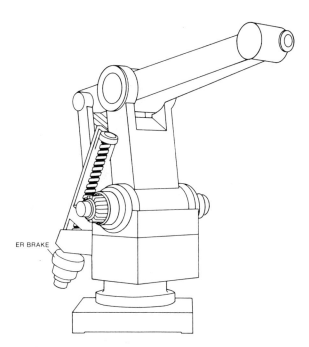

ER BRAKE

FIGURE 6.24. Ball bearing screws used in the drive systems of jointed-arm robots. Courtesy of Warner Clutch and Brake Co., South Beloit, Ill.

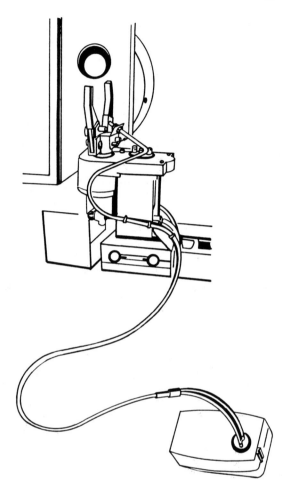

FIGURE 6.25. Gripper monitoring. Courtesy of Warner Electric Brake and Clutch Co., South Beloit, Ill.

1. To "lock" the manipulator arm into position in order to hold a payload at a desired point in the operations sequence.
2. To prevent the manipulator arm from dropping or swinging in the event of power failure. This protects the robot, tooling, fixtures, and its payload from damage.

ER brakes can be applied on every axis of the robot. A common configuration is to mount the brake between a dc servomotor and a ball bearing screw.

6.13   Photoelectrics and Robotics

Robotics today is at the beginning of a transition from first-generation to second-generation technology. Most robots currently in use are first-generation machines. Although they can be reprogrammed for a variety of tasks, they are essentially fixed-sequence machines. Each workpiece must be in the right place at the right time with the proper orientation.

The second-generation technology makes the robot more adaptive by giving it some kinds of sensing capabilities. Adaptive robots will be able to deal with variations in the planned operation (e.g., a part coming down a conveyor in an assembly operation that is misoriented or out of sequence).

Computer "vision" systems are discussed in Chapter 7 and various "touch" schemes have received most of the attention given to robot sensing. It is generally believed that touch or tactile sensing technology will require many more years of development before it is ready for general industrial applications. Several computer-controlled robot vision systems are on the market today, although this area of technology also has a long way to go before robots can really approximate "sight" in the customary sense of the word. Computer vision systems currently available can be used for gaging, sorting, positioning, orienting, inspection, part present verification, and the like.

Photoelectric sensors (discussed in Chapter 5) fulfill several of the same sensing functions in a robot operation in a much simpler manner at a much lower cost. Some examples include the following:

1. *Part present/absent detection.* Photoelectric sensors monitor a fixture to assure that all parts necessary for the next sequence in the operation are present before the robot begins this sequence.
2. *Gripper monitoring.* Photoelectric sensors monitor a gripper to assure that a part has been picked up or to assure that a part has been released before another part is grasped (see Figure 6.26).
3. *Actuation of gripper.* In some cases, a photoelectric sensor can be used to detect the presence of an object between the gripper and signal the gripper to close at the proper time.

Fiberoptic photoelectrics are especially well suited to robot applications. The small sensing head can even be mounted directly on the gripper, with the control box mounted nearby where more room is available. Other advantages of fiber optics include the following:

1. Flexible cable can move with the robot arm.
2. High tolerance of vibration.
3. High tolerance of wide temperature range.
4. Relatively immune to damage from fluids.

Robot end effector actuation is monitored and controlled by LED photoelectric scanners. The rugged, compact sensing heads of a Warner Electric fiber optic scanner can be mounted directly to the gripper, and remotely connected to control electronics by flexible cables.

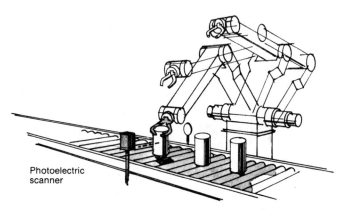

Photoelectric scanner

Part present/absent detection at the presentation station is one of a variety of photoelectric sensing functions. Other robot peripheral applications for Warner Electric photoelectric scanners include sorting, positioning and orienting workpieces or payloads.

FIGURE 6.26. Gripper monitoring with a photoelectric scanner. Courtesy of Warner Electric Brake and Clutch Co., South Beloit, Ill.

6.14 Harmonic Drive Transmission System and Gearing Installation Used in Robotics[e]

A harmonic drive transmission system (Figure 6.27) operates on a patented principle that uses a deflection wave transmitted to a nonrigid Flexspline member to produce a high mechanical advantage between concentric parts. The three elements shown in Figure 6.27 are typical and common to all harmonic drive units and function in the following manner.

The Flexspline assumes an elliptical shape upon insertion of the wave generator into the bore. The resultant spline pitch diameter at the major axis becomes the same as the circular spline teeth at two points 180° apart to form a positive gear mesh.

One clockwise (CW) input revolution of the wave generator produces one CW revolution of the Flexspline elliptical shape, causing a continuous tooth-to-tooth rolling mesh at the two points of engagement. The resultant rotation of the Flexspline proper relative to a fixed circular spline is two teeth in a CCW direction or a reduction ratio equal to one half the number of teeth on the Flexspline.

The harmonic drive transmission system is used in robots because of its precise angular positioning characteristics with superior weight to torque ratio. The standard harmonic drive will show essentially zero backlash and

Circular Spline
A rigid thick wall ring with internal spline teeth. It is a fixed or rotating output drive element.

Wave Generator
An elliptical ball bearing assembly which includes an oldham type shaft coupling. It is the rotating input drive element.

Flexspline
A non-rigid cylindrical thin wall cup with two less spline teeth and on a smaller pitch diameter than the circular spline. It is a fixed or rotating output drive element.

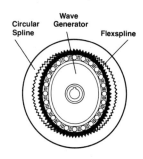

Circular Spline   Wave Generator   Flexspline

FIGURE 6.27. A harmonic drive transmission system. Courtesy of Harmonic Drive Division, Emhart Machinery Group, Wakefield, Mass.

---

[e] Section 6.14 is from *Designing with Harmonic Drive,* © 1982, and is used courtesy of the Harmonic Drive Division, Emhart Machinery Group, Wakefield, Mass.

cannot be differentiated from normal torsional system windup. The effect of tooth-to-tooth error is also minimized with harmonic drive. This is due to the fact that approximately 10 % of the number of teeth are in continuous engagement. This contributes to accuracy capabilities in the arc minute range with repeatability in the arc second range.

The harmonic drive pancake gearing installation (Figure 6.28) is also used in the mechanical system of robots. Such gearing systems offer high ratio, in-line mechanical power transmissions in an extremely compact configuration. The gearing component consists of four elements: the wave generator or input element, an elliptical bearing; the Flexspline, an elastic ring of steel with external teeth; the circular spline and the dynamic spline, rigid steel rings with internal teeth. These types of drives, however, exhibit some pure backlash in the arc minute range.

6.15   Shock Absorbers Used in Robotics[f]

Actuators (discussed in Chapter 4) often require external absorbing systems to reduce robotic machinery shock loading to prevent the instrument from failing. In robotics, the shock absorbers can provide three-axis deceleration and positioning for pick-and-place robots. Enidine, Inc., of Buffalo, New York, has specifications for such shock absorbers, which includes the following:

1. Stroke: 0.38 in.
2. Maximum energy per cycle: 30 in.-lb
3. Maximum shock force: 80 lb
4. Maximum propelling force: 80 lb
5. Weight: 2.1 oz

A photo of the Enidine shock absorber used in robotics is shown in Figure 6.29.

Motion applied to the shock absorber piston pressurizes the fluid within the device and forces it to flow through restricting orifices, causing a heat rise in the fluid. Damping is thus accomplished by converting the vertical motion energy (kinetic energy) into heat (thermal energy), which is then dissipated via the cylinder to the atmosphere.

This process is demonstrative of the basic principle of the conservation and transformation of energy: Whenever a body does work, the capability to do further work is reduced by the amount of energy imparted to the body or medium. By this same principle, the energy thus transferred or transformed is not destroyed but simply changed.

---

[f] Section 6.15 is from *Controlling Motion Effects: Its Principles and General Applications*, Technical Bulletin 3000, and is used courtesy of Enidine, Inc., Buffalo, N.Y.

TYPICAL INSTALLATION

HDUF Pancake Gearing Component Sets are easier to use than conventional gearing. All that is required is suitable bearing support of an input and output shaft, and a means of fixing the Circular Spline against rotation.

The simplicity of HDUF Pancake Gearing is demonstrated in the typical arrangements shown below.

1. Wave Generator
2. Flexspline
3. Circular Spline
4. Dynamic Spline
5. Motor
6. Input Shaft or
   Motor Shaft
7. Output Shaft

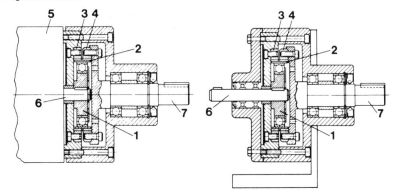

FIGURE 6.28. Typical HDUF pancake gearing component sets. Courtesy of Harmonic Drive Division, Emhart Machinery Group, Wakefield, Mass.

Such energy dissipation is a one-way process; there is no inherent energy storage, no capacity for effecting piston return. Almost all work processes, however, require cycling. There is a need to restore the device to its original position so that it may perform again, and the rate of return may bear heavily on productivity. The optimum relationship between energy dissipation and restoration rate is achieved in Enidine shock absorber systems.

FIGURE 6.29. Shock absorbers for robotics. Courtesy of Enidine, Inc., Buffalo, N.Y.

Through the use of an elastomeric urethane, it is possible to achieve high reliability and zero leakage in a device to control motion effects. This approach also permits a significant miniaturization; units have been produced as small as 0.25 in. diameter by 1.00 in. long. When combined with contamination-free materials of construction, they offer an attractive answer for applications in which oil or other contaminants are prohibited. Its limitations are found principally in the limited operating temperature range (70°F ± 25°F), relatively small size potential, and restriction to low cyclic operations. The small nonliquid device placed into a closed container with a plunger has the following characteristics:

1. Energy: 0.5 to 1400 in.-lb
2. Stroke: 0.10 to 1.5 in.
3. Temperature: 70°F ± 25°F
4. Weight: 1 to 20 oz.

Such a device can be reused in industrial robotics.

6.16   Review Questions

1 . Discuss end-effectors used in robotics.
2. Discuss the use of grippers in robotics.
3. Discuss transporters in robotics.
4. Discuss rotators in robotics.
5. Discuss the cable drive system in a robot system.
6. Discuss the ball screws and spline in robot systems.
7. Discuss the use of the electrically released type of brake for robots.
8. Discuss photoelectric sensors used with robots.
9. Discuss the harmonic drive system used in industrial robotics.
10. Discuss the need for shock absorbers in industrial robotics.

6.17   Bibliography

J. Duffy, *Analysis of Mechanisms and Robot Manipulators,* John Wiley & Sons, Inc., New York, 1980.

C. S. George Lee, "Robot Arm Kinematics and Control," 0018–9162/82/1200–0062500.75, *IEEE Computer,* Dec. 1982, pp. 62–79.

Microbot Gripper, A publication for users of Microbot industrial and educational robot arm, Vol. 1, No. 1, Oct.–Dec. 1982, published bimonthly by Microbot, Inc., 453-H Ravendale Drive, Mountain View, Calif. 94043.

Janice Mullins, "Ball Bearing Screw Actuators," *Power Transmission Design*, Dec. 1982.

R. P. Paul, *Robot Manipulators: Mathematics, Programming, and Control*, MIT Press, Cambridge, Mass., 1982.

Mark Rosheim, "The 2–Roll Gripper," *Robotic Age*, Jan.–Feb. 1983.

Harrison Thomas, *Introduction to Kinematics*, Reston Publishing Co., Reston, Va., 1976.

Martin Weinstein, *Android Design*, Hayden Book Co., Inc., Rochelle Park, N.J., 1981.

7

Robot Vision Systems

1.1   Introduction

Robots perform complex industrial tasks such as inspection and assembly. Industries use robot vision to enhance the abilities of the machinery in the factory.

A robot vision system uses a sensor sensitive to light, temperature, radiation, and the like, that transmits a signal to a measuring or control device. The robot vision system actually has an input device such as a vidicon camera sensor or sensors or a solid-state diode array. The input video signal has been digitized via a data acquisition and conversion system; the signal is transferred to a computer as an image. A computer must treat the visual image as an array of brightness dots called picture elements, or pixels. A typical scene may consist of from 16,000 to over 1 million pixels. Interpretation of such a large volume of data is an enormous task even for a high-speed computer. It often takes many seconds to several minutes to analyze a single picture by the computer. This is far too slow for the robot to respond in a timely fashion to what it sees. Various tricks are used to speed up this response time. One is to illuminate the scene so that the objects appear as black and white silhouettes. Another is to assure that no two objects of interest touch or overlap. However, even under such artificial circumstances, robot vision is a very complex problem and subject to many difficulties.

Arrays of photosensitive elements are also available in a wide variety of shapes and sizes. As an example, consider a moving belt conveyor. Linear arrays mounted over the conveyor have resolution to over 2000 elements.

In stereo vision systems, two cameras are located at different points to extract the information from the scene.

One difficult robot vision problem is referred to as "bin-picking," that is, retrieving objects that are jumbled together in a bin, the way parts are usually delivered to the factory area.

Vision systems in industry are generally used for the following functions:

1. Identification
2. Classification
3. Sorting and measuring
4. Inspection
5. Verification
6. Quality control
7. Forging turbine blades

As a simple example, suppose we wish to watch a robot playing a game of chess with someone. We could use a Microbot Mini Mover-5 robot arm, a solid-state camera and two Z80 microprocessors or the equivalent, as shown in Figure 7.1.

The camera shown in Figure 7.1 is placed above the subject or the robot. In many industrial applications, the camera is attached to the top of the robot. In the near future one can foresee the camera inside the robot arm or as a part of the software equipment.

The Pixelcaster digital processing system (Figure 7.2) is used in robotics for the following:

1. Image and feature recognition
2. Noncontact measurements
3. Go–no go decision making
4. Image feedback
5. Task orientation

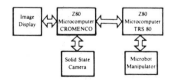

FIGURE 7.1. Block diagram of a robot vision system. Courtesy of Microbot, Inc., Menlo Park, Calif.

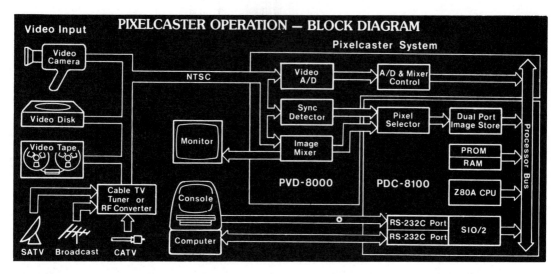

FIGURE 7.2. Block diagram of Pixelcaster operation. Courtesy of Vision Peripherals, Anaheim, Calif.

The Pixelcaster can be used for processes like part positioning, automated assembly, template matching, bin-picking, target classification, spot welding, arc welding, and machine loading and unloading.

The Pixelcaster (Figure 7.2) interfaces via RS–232–C and accepts ASCII commands. It features a Z80A CPU and a high-speed memory for storage of up to 1024 image samples and coordinates of the sample locations. Sample timing is referenced to the NTSC 3.579545–MHz color subcarrier, which makes for very stable sampling of TV images. The 1024 four-bit pixels can each come from anywhere in the TV signal, including the sync intervals or even consecutive fields. The order of the pixels transmitted to the host in ASCII is independent of the order the samples are acquired in during the TV raster scan. Special X and Y offset registers allow the specified pattern of pixels to be quickly translated about on the image, and real-time interrupts to the Z80A CPU enable processing to be synchronized to events in the video signal.

Under program control, the 4 bits in a sample can encode high-resolution samples (910 per TV line), 4 bits of amplitude information, or a mixed mode including edge detection. The 4-bit amplitude digitizing is done within a "window" set by two 8-bit DACs, so amplitude resolution of 1 part in 4096 is available. Monitor output signals allow programmable viewing of the incoming video, 1-bit processed video, 4-bit digitized video, and the sample pattern, either in color or black and white.

A 1–Kbyte RAM is available onboard for data and code, along with 2K or 4K of ROM. An additional 48K of RAM and up to 16K of ROM can be added simply by plugging in the PME–8052 Memory Expansion Board. User-defined Z80 code modules can be developed as part of the local preprocessing. When a user-defined routine is loaded and executed, all the capabilities of the hardware are accessible.

The Pixelcaster can be configured as a two, three, or four board set. Although they can stand alone for special applications, the PDC–8100 Digital Controller Board and PVD–8000 Video Interface Board together make up a powerful 1-bit, 7-MHz intelligent video preprocessor. Plugging in the optional PVE–8800 Video Expansion Board adds 4-bit, 14-MHz sampling, dual 8-bit window-setting DACs, four selectable video inputs, an onboard sync generator, and dual 8-bit output ports. Rounding out the set is the PME–8052 Memory Expansion Board, although Vision Peripherals will soon be adding a Frame Store and Video Convolver to the family. All the boards are S–100 size, but only power is drawn from the bus. RS–232–C is the only I/O required, which greatly simplifies the use of the system and allows remoting of Pixelcasters from a host computer.

The Univision I™ robot vision system (Figure 7.3) consists of a vision processor, graphic pen, camera(s), and Unimations VAL software and hardware interface (discussed in Chapter 8).

The automated system for the factory floor interfaces with PUMA robots to improve inspection, parts sorting, and automated interface.

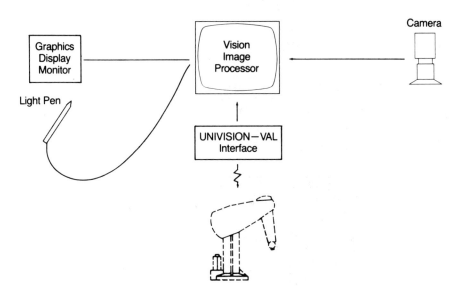

FIGURE 7.3. Univision I system concept. Courtesy of Unimation® Inc., Danbury, Conn.

The UNIMATION II™ robot vision system is designed for robotic arc welding, expanding the ability where seam displacement deviates from one weldment to the next (see Figure 7.4).

The Robotic Vision Systems, Inc., of Melville, New York, has developed a three-dimensional electro-optical Robo Sensor™ Model 200 Series vision system that rapidly and accurately measures the position, orientation, width, and depth of the seam to be welded; this truly represents the most advanced and most reliable vision equipment available today to the welding industry to meet its present and future expanding needs.

Basically, the Robo Sensor™ vision system, which is comprised of a rug-gedized vision module mounted on the robot's arm and an electronics cabinet that includes a microprocessor, looks at the surface of the seam and sends out

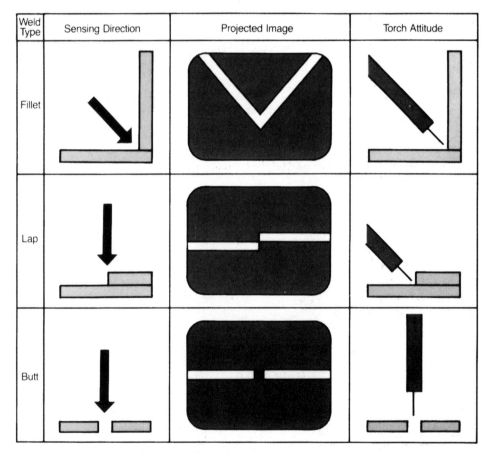

Weld Type	Sensing Direction	Projected Image	Torch Attitude
Fillet			
Lap			
Butt			

FIGURE 7.4. Univision II system concept. Courtesy of Unimation® Inc., Danbury, Conn.

a digital data stream that accurately defines the seam's position, orientation, width, and depth.

Although the Robo Sensor 3–D vision module is configured for mounting on the tool mounting plate of a Cincinnati Milacron T_3 Industrial Robot, the vision module can also be configured for use with other robots or mechanical translation equipment.

Electro-Optic Integrated Systems of Carlsbad, California, has a robotic vision system called the Vobot. It is composed of 13 boards, a capacitor charged device (CCD) camera, computer, and camera interfacing circuits. The Vobots perform the 13 most common functions for machine vision, including distance, position, angle, gray-scale, and orientation measurements.

Table 7.1 provides insight into some of the available commercial robotic machine vision systems. Tomorrow there will be many more such systems.

In the sections that follow, we shall examine the general machine vision concept and an application for inspection in an automated factory. There are many industrial manufacturers making robot vision systems. Hundreds of applications are possible and newer concepts will be forthcoming from academic sources and industry.

TABLE 7.1
SOME COMMERCIAL ROBOTIC MACHINE VISION SYSTEMS

Company	Model	Gray-scale levels [black and white]	Resolution [pixels]	Processing speed
1. Automatix Billerica, Mass.	Autovision II	16	256 × 256	360 parts per minute
	Autovision IV *	256	512 × 256	1800 parts per minute
2. Unimation Danbury, Conn.	Univision I	2	256 × 256	Less than 500 milliseconds per frame
	Univision II	2	256 × 240	56 milliseconds per frame
3. Vision Peripherals Anaheim, Calif.	Pixelcaster	16	1024	
4. Octek, Inc. Burlington, Mass.	20/20 Vision Development Series	16	320 × 240	300 parts per minute
5. General Electric Syracuse, N.Y.	Optamation II	12	244 × 248	
6. Electro-Optic Integrated Systems Carlsbad, Calif.	Vobots			

* Camera accommodation: 16

7.2　　Machine Vision in Robotic Systems[a]

The six basic areas of a robot vision system for industrial machine vision are the following:

1. Lighting and lensing
2. Video imaging
3. Video image digitizer
4. Window comparator
5. Machine control logic
6. Operator interfacing

Gray-scale processing can be combined with a window comparator and pixel counting as in digital signal processing to perform high-speed image comparison.

Like photography, the key to a good machine vision system application lies with the proper selection of lighting and lensing elements. The highest contrast between individual features on production parts will allow the most consistent and reliable operation.

The requirement for optical applications assistance will grow as speeds increase and synchronization becomes more important as production lines tend toward 100% in-line inspection. In many cases, customers rely on outside assistance although they supply their own mechanical fixturing and electrical/electric interfacing.

Many systems can be used to create a video image, with the most common method employing a Vidicon black and white camera, similar to those used with home video recorders. Variations of this camera include the Newvicon used for nighttime parking lot surveillance and the Ultracon used for more specialized applications. These units have proven reliability and significantly reduced prices due to multimillion unit production runs.

Machine vision systems use video image digitizers to benefit both concerns because they simplify the image and look only at features relevant to the process. Consequently, most systems use black and white images, rather than color, and convert the information into black and white dots to eliminate intermediate gray levels. This conversion occurs repetitively and is best performed in high-speed dedicated electronic logic. Such hardware processing is especially critical when machine vision is used for 100% in-line automated environments.

Typically, a video image is captured every 16 milliseconds and sent to the initial processor, a video image digitizer. Each image is electronically frozen

[a] Section 7.2 is reprinted from Newell V. Starks, Jr. "Robotics and Machine Vision in the Electronics Industry," *Test and Measurement World*, Oct., 1982, and is used courtesy of the Interfeld Publishing Company, Inc., Boston, Mass., © Oct. 1982.

and divided into a grid of 254 vertical by 254 horizontal lines yielding 64,516 picture elements, or pixels. The number of pixels in the horizontal or vertical image determines system resolution. For example, consider a 10-in. printed circuit board (PCB) with hole dimension tolerances of ± 0.001 in. If the entire board were viewed at once with a system comprising 254 pixels, each pixel would contain 10 in. divided by 254 or 0.039 in. of PCB. This is inadequate resolution. However, by using lensing to magnify the image and reduce the field of view, an area of 0.100 in. could be viewed at one time. In this case, each pixel would contain 0.100 in. divided by 254, or 0.00039 in. This is sufficient resolution for measurements of ± 0.001 in.

There is always a tradeoff made between resolution required and field of view available. When an application requires a resolution better than one part in 254, multiple cameras spaced across overlapping fields of view can improve the overall resolution. Also, mechanical fixturing such as X–Y stages can step and repeat portions of the entire part (called "cells") in front of a highly magnified camera.

Window comparators provide a powerful process for feature analysis. As an example, window comparators provide a technique to perform inspection, sorting, and rudimentary positioning. Similar to a window in a wall, an electronic window restricts what a computer "sees" of an image. By positioning each window over an area of interest from a known good part, such as a hole, stamped, machined, cast, or extruded feature, the machine vision has a benchmark to compare production part features.

For example, in Figure 7.5, if three component dimensions, A, B and C, need to be measured, a machine vision system can "draw" three windows, each a pixel wide and long enough to clearly pass over the entire length to be measured. Each window forces the computer to ignore all other elements of the image by masking out extraneous portions. The system then merely counts the number of black or white pixels in each window to measure each dimension in terms of pixels. Since each pixel has a known size, which can be easily determined during calibration, a mathematical ratio applied to the pixel count yields a precise dimension.

When performed at high speed and for many features, this simple count of pixels can yield extraordinary results in manufacturing processes. At rates up to 30 parts/s, a machine vision system can measure line widths on a hybrid circuit trace, check for visible shorts in a thick film process, examine the position of a component, or verify a fiducial alignment mark—all simultaneously.

Micro- or minicomputer systems typically cannot process image data quickly enough to give real-time feedback on a variety of features. Dedicated electronic hardware can perform window-to-image comparisons, allowing the processing of 1000 windows 30 times per second, the equivalent of examining 30,000 features each second.

The capability to make decisions and to perform real-time machine control logic is essential for moving machine vision out of the laboratory and into

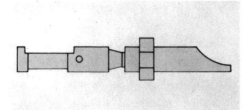

a. In this example, a connector pin is to undergo inspection

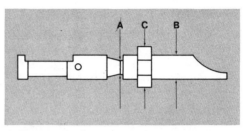

b. Three areas of critical measurements are established.

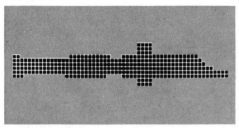

c. The connector pin is digitized.

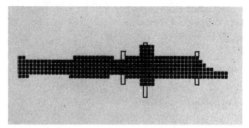

d. Measurement windows are placed over the digitized pin.

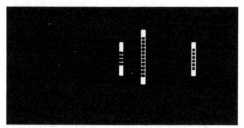

e. Pixels are isolated within each window.

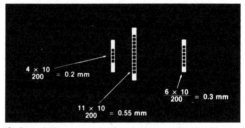

f. Actual sizes are calculated and compared with stored data.

FIGURE 7.5. Connector pin inspection steps. Courtesy of the Interfeld Publishing Co., Boston, Mass., © 1982.

the production environment. This will allow independent automated processing for achieving 100% in-line inspection. The need to control machinery is especially evident when a vision system must drive an X–Y table to perform its full operation.

Automatic semiconductor bonding machines are a classic example of a fully automated, visual feedback system. Typically, the X–Y stage of the bonder grossly locates the semiconductor die under the field of view. The machine vision system then looks at the bonding pad with respect to a fixed or relative coordinate system and calculates the mechanical offset to create a centered bond. It then, at very high processing rates, corrects the X–Y table, activates the bonder, and proceeds to the next bond.

With a choice of industry standard bus structures, vision systems can support a wide array of industrial control interfaces. These interfaces enable the system to perform wide-ranging tasks based on complex visual decisions to suit production requirements and stop errors before they occur.

For machine vision systems to succeed as a production tool, the operator must be able to interact with the tool in a straightforward, uncomplicated way.

Some insight into the operator interface might be gained from robotic mechanism installations. In numerous cases, the success or failure of a robot installation depends on the attitude and acceptance of production floor personnel. In the attempt to help improve throughput and product quality with machine vision, we must not forget this keystone to installation success.

Fortunately, advancing technologies provide us not only with the capability to create better tools, but with the ability to make them intuitively obvious to use. With machine vision, for example, the part under inspection may be displayed on a cathode-ray tube to help orient the operator; furthermore, an interactive menu can be displayed on the same monitor to lead operators through the proper procedures and to prompt them when necessary.

Systems will have to address a wide range of potential operators. First, there is the actual production operator; second, the model changeover person; third, the new model setup person; fourth, the manufacturing engineer who designs the system; and fifth, the technical specialists who may be required during system design. By considering each level of operator, we can design interfaces that are easy to use and provide the flexibility to allow more involved access during design/installation.

Applications for robotic machine vision in automated production lines using visual feedback also holds for the following:

1. Silk screened ceramic substrates
2. Hybrid component placement
3. Integrated circuit conformity
4. Rotating shafts

For the next two or three years, most machine vision applications will evolve logically from current automation practices such as mechanical fixturing of parts. In that time, sophisticated algorithms will provide the speed, cost reductions, and precision needed to allow fewer constraints in mechanical fixturing.

Consequently, in-line visual inspection for robotic systems will focus on high-volume products such as integrated circuit bonding and substrate component placement verification. As systems gain more power and users acquire more applications knowledge, vision systems will inspect complex subassemblies, such as printed circuit boards, component types, lead placement, and solder shorts. Ultimately, systems could examine the interior of a chassis for scratched paint or incorrect cable routing.

Automated robotic machine vision is already finding use in inspecting bare printed circuit boards for defects, dimensional inspection, stuffed board inspection, and PCB artwork and digitization. Other areas involved in the PCB production area include from inspection of board drilling to the inspection of imaging, etching, and solder coating. Automated printed circuit board inspection lowers labor costs, reduces scrap and error rates, and improves uniformity of inspection results. In other words, the quality control of the product is improved.

7.3   Robotic Vision Application System for Automated Manufacturing[b]

Inspection and measurement applications can be coupled to robots and other automated manufacturing system controllers to provide a range of robotic vision functions.

The vision systems offer the high resolution, geometric stability, and image enhancement features needed for proper inspection and measurement applications. They can be programmed to detect incoming parts, to pick out key part features for verification or recognition, to provide part features for verification or recognition, to provide system guidance (for example, seam tracking for automated welding), or to perform other automatic alignment functions in robot or dedicated automated manufacturing. Communication to the robot controller is typically handled through an RS–232–C link.

Applications include inspecting critical part dimensions, position measurement or alignment, gap location and measurement, seam tracking, defect detection, and acceptance testing. The systems can be used as stand-alone inspection systems or can perform vision functions in addition to the measurement or inspection.

Dimensional measurement functions and defect inspection functions frequently can overlap with one system performing recognition, measurement, and inspection functions. The dimensional functions are more easily generalized and universal. The system must be set up for the appropriate part size, measurement speed, part complexity, the number of dimensions to be measured, and requirements to ignore or to quantify localized defects and dimensional variations.

Inspection systems to identify and classify defects must be programmed for specific defect characteristics. Our basic approach is to match an existing visual inspection procedure to define the defects or characteristics of interest and then to implement appropriate algorithms to automate the inspection process.

---

[b] The material in Section 7.3 is used courtesy of Spectron Engineering, Denver, Colo.

The systems are modular to adapt to diverse applications. The imaging portion consists of cameras using linear or matrix photodiode arrays (or other solid-state imagers) in a range of sizes. Various optical front-ends cover a wide range of field-of-view and working distance requirements. Typically, a low-power pulsed xenon light source is used. Its very short duration allows sub-microsecond image acquisition, eliminating vibration or motion as error factors. Steady-state or laser sources are used if they are appropriate.

The microprocessor-based data analyzer and system controller performs signal conditioning and correction, image processing and enhancement, data analysis and extraction of relevant dimensons or features, pass/fail analysis in inspection applications, and communication to an operator or to other systems such as a robot controller. Multiple cameras in several locations in a work area may be handled by one controller.

These vision systems are factory programmed to perform specific functions. Optional "learn" modes allow the system to pick up key features or dimensions from standard parts, nominal or limit samples, as a basis for acceptance or rejection.

The photodiode array detectors are not subject to geometric and signal-amplitude distortions and instabilities that limit the use of vidicon-based systems for dimensional measurement and defect detection. While they may be suitable for recognition or position/pose functions, vidicon systems are generally not suitable for the frequently coincident measurement and inspection requirements.

The image-enhancement algorithms use the full gray-scale information available in the image to give resolution to a small fraction of the pixel-to-pixel spacing, rather than the several-pixel resolution typical of systems that "simplify" the image by reducing it to a binary (black and white) image. Typical resolution improvement is by a factor of 100.

There are two specific problem areas where the Spector vision and inspection system will aid the automatic robot insertion process. The first relates to the inspection of the component leads to assure that all eight leads are within an acceptable tolerance band so that they will properly insert into the PCB hole pattern. This inspection should greatly reduce the number of components destroyed by the insertion process and should therefore yield significant economic benefit in part cost saving and should also allow a longer production run with the existing parts, another major economic factor.

The second area is in the measurement of the board hole pattern position of the circuit board. The variability of the component lead hole pattern in the circuit board relative to the positioning holes takes up a significant portion of the available component lead tolerance. This variability of the boards will significantly reduce the allowable tolerance for the component leads, giving increased rejects for hand insertion and increased component and board scrap due to smashed component leads. In fact, the combined variability of the circuit boards and the tolerance of the robot arm position for insertion may give no worst-case tolerance for the component lead positions. It is clear that some means of shifting the circuit board or the robot is desirable to increase the yield and reduce the scrap.

One added area addressed (and covered as a system option) is to inspect the parts and to adapt the circuit board or the robot position to correctly insert the component if possible even if individual leads were out of tolerance.

In any case, rejected parts will be dropped into an appropriate bin for hand insertion. The system will automatically provide a warning if no part was picked up for insertion.

Although the robot coupled vision and inspection system provides several specific vision functions for the robot component insertion system, it is based upon and shares most components with the noncontact inspection and measurement systems.

The system uses modular components, including the CE400 microprocessor-based system controller and data analyzer, one or two CE410 line scan cameras (one modified for two-dimensional position measurement), light sources, and the necessary interface to couple to the robot system.

The CE400 controls the operation of the system. Its functions include communication with the robot controller; control of the test timing and related test conditions; control of the (CE410) camera operation; A/D conversion, noise subtraction and signal conditioning of the analog signal from the camera; interpretation and enhancement of the camera signal to develop the required position data; and comparison of the position to the present position tolerances. The controller will have additional functions such as computing the required coordinate position shifts or actually controlling the dc motors to shift the PCB holder stage if those optional functions are incorporated. The CE400 also provides the power and timing control for the pulsed (strobe) light sources used to backlight the component leads or frontlight the PCB.

The CE400 controls and monitors the overall system operation, measuring focus and lighting, as well as timing. It can provide warnings if these are outside of prescribed operating parameters so that an operator or supervisor can check and correct any problems.

The CE400 is based on a Commodore 6502, 8-bit microprocessor. It has provision for a significant amount of both RAM and ROM. The operating programs are written in machine language. Any operating system modifications can be reprogrammed at our facility. Changes that only affect calculations or numerical tolerance factors can be supplied as plug-in replacement EPROMs or plug-in circuit cards. A change, for example, in the tolerance dimension for the lead could be easily implemented.

The CE410 line scan camera is an electronic camera based on a linear self-scanned photodiode array. For the lead position measurement application, the camera is aligned so that the image formed is perpendicular to the lead to be measured. For this application, the lead is backlighted so the image will consist of a shadow with soft edges on a bright background. A dual imaging optical front end is used so that the camera simultaneously images an X and a Y view onto two portions of the same array. The two portions of the array will effectively be analyzed separately to give an XY position. The signal is analyzed in the CE400, and the position measurements are developed and compared to the tolerances in the controller. The optical front end is a custom feature for this application.

A second camera is coupled to the same controller if the circuit board position function is included in the system. The second camera will actually not use the standard linear photodiode array, which gives only a single line image. It is best for this application to use a small matrix array that also fits into our CE410 camera structure. The camera is set up so that it images across component holes so that we can identify X, Y, and rotational displacement. The camera is also equipped with a long working distance lens so that there will be no interference with the robot arm.

The basic system function is to inspect each of the eight leads on each component prior to robot insertion of the component into the printed circuit board. The part handling for the inspection step will be done by the robot as part of the sequence of picking up the part, cutting off the leads and inserting the component into the correct hole pattern on the board.

Specifically, the steps to cut off the leads to the correct length would be completed, and then the robot would bring the part into the field of view of the inspection system. The system would wait until it received a ready pulse from the robot controller indicating that the lead was in position for the measurement. There would be a fixed delay for the measurement to be completed and then the robot would rotate the component to position to measure the next lead. The process would continue for each of the eight leads. The inspection system would provide a reject signal if any of the leads was out of the established tolerances. On a reject signal, the robot would drop the part in a bin to separate parts for hand insertion. If there was no reject signal, the component would be inserted into the printed circuit board.

An added benefit of the inspection systems is that it can test for missing components. A routine is incorporated that provides a separate output if no leads were present because this would indicate a missing part. From the demonstration of the robot operation, it is possible to get no component if the holder sticks in place (leaving an empty holder) or if the holder was too tight so that the component was not picked up. A separate latched TTL output is provided that will warn the operator if two consecutive cycles show no leads. This output could be used to trigger an audible or visible alarm to bring an operator to correct the problem.

The actual XY position of each individual lead will be compared to a predefined position tolerance. The tolerance defines a maximum displacement from the nominal position or, effectively, a circular area of acceptance. The tolerance will be programmed in EPROM rather than keyboard entered. A change in the tolerance would be accomplished through a change in the

EPROM. It is necessary to confirm the position of each lead relative to the robot. The robot will do the insertion based upon its predefined insertion position; if the component is somehow shifted with respect to the anticipated position, it will not be properly inserted even if the component lead spacing would otherwise allow insertion. Therefore, measurement of the lead to lead spacing is not needed. In a later stage of this quotation, we discuss a more sophisticated system that either shifts the robot coordinate frame or the PCB to adapt to the component.

The system will be provided with an automatic power-up reset. The system will automatically come up in the operating mode for inspecting the lead positions. Since the position tolerances are defined in EPROM, the operation would be unchanged after a power glitch. The system is less subject to operator setup errors if the tolerances cannot be changed. However, if the tolerances are to be operator controllable, an advantage during the initial setup phases of the work, and retain the automatic restart of operation after a power interruption, one would add a nonvolatile RAM board to the CE400 and have the lead tolerance enterable through the keyboard.

The dual axis camera will use mirrors or prisms to simultaneously image from two axes so that the XY position is computed.

The variability in the position of the component pads and holes in the printed circuit boards is compared to the alignment holes. The board alignment holes control the position, but the robot coordinate reference frame and therefore the actual insertion is based on a nominal position that may be off (by about ± 0.006 in.). This significantly decreases the allowable tolerance for the components. As a worst case, if we include the tolerance of the robot and the board position, we may not have any component lead position tolerance.

This second level of the vision system functions will allow for either shifting the robot coordinate reference frame or translating the circuit board to eliminate this variability.

A second camera is added to the system along with analysis programs that will determine the board position, and compare it to the nominal and output correction factors. There are two methods of achieving the correction. If it is a simple procedure to shift the robot's coordinate reference frame, that would be the best approach. The output will have three 8-bit words giving the X, Y, and rotational offset. The words will be latched TTL outputs. If it is easier to shift the board, motor drive outputs will be used to control the X, Y, and rotation of stages provided to shift the circuit board. These outputs will be analog timed outputs to drive dc motors.

A small matrix camera is used for this function with the camera and light source mounted above the normal operating positions of the robot arm. It may be necessary for the arm to move out of the way briefly while the image is acquired, but since this step is done only once per board, the small time delay should not be a factor. The displacement can be determined by looking at any two holes on the board. The two holes toward the top center and bottom center of the board are imaged, although the precise holes may change. The image will be analyzed in the CE400 with the actual positions compared to the nominal position. The CE400 controller will then either provide the output to the robot or will drive the stages to move the circuit board.

The basic component lead tested determines that each lead is within a defined area so that it would properly insert. It is possible for the entire set of eight leads to be shifted in the X or Y axis or rotated. Some components therefore could be robot inserted only if shifted. They would be rejected and loaded into the hand insertion bins unless there was some way of identifying the necessary shift and of implementing it.

The inspection system calculates an appropriate shift by storing the location of all eight leads and then determining what shift, if any, would make the part properly insert. The advantage of this more sophisticated approach is that it would allow for automatic insertion of some components that otherwise would be put in a bin for hand insertion.

This function requires significantly more computation and therefore more complex programs in the CE400 controller. If the percentage of parts automatically inserted increased significantly with this feature, it would be worthwhile to implement.

Solid-state image sensors are also used in robot vision for the movement and control of robots. Assembling, sorting, welding, painting, and other production line operations use robots where recognition of distinct patterns and shapes is required. Such sensors are adaptable for the factory of tomorrow.

7.4   Review Questions

1. Discuss the fundamentals of a robot vision system.
2. Describe a robotic machine vision system.
3. Discuss the application of a robotic inspection and measurement system in an automated factory.

7.5   Bibliography

King-sun Fu, "Pattern Recognition for Automatic Visual Inspection," 0018–9162/82/1200–00345.75, *IEEE, Computer,* Dec. 1982.

Rafael C. Gonzalez and Reze Safabakhsh, "Computer Vision Techniques for Industrial Applications and Robot Control," 0018–9162/82/1200–00175000.75, *IEEE Computer,* Dec. 1982.

John F. Jarvis, "Research Directions in Industrial Machine Vision: A Workshop Summary," 0018–9162/82/1200–0055500.75, *IEEE, Computer,* Dec. 1982.

Paul Kinnucan, "Machines That See," *High Technology*, P.O. Box 2808, Boulder, Colorado, 80321, © 1983.

Proceedings of the 2nd International Conference on Robot Vision and Sensory Controls, International Society for Optical Engineering, SPIE, Volume 392, Stuttgart, Germany, Nov. 2–4, 1982. Published by IFS (Publications) Ltd. 35–39 High Street, Kempston, Bedford, England, Nov. 1982.

Edward L. Stafford, Jr., *Handbook of Advanced Robotics*, Tab Publishers, Summit, Pa., 1982.

Vision for Automatic Inspection and Robotics, Session 2D, Electro '83, presented at the Sheraton Center, New York, April 21, 1983.

8

Practical Industrial Robots: Specifications and Applications

8.1   Introduction

In this chapter, the authors will explore a practical industrial robot system used by a leading manufacturer, but the information contained is applicable to many industrial robotic systems. In addition, thumbnail sketches including specifications and applications are given for three of the leading industrial robotic manufacturers in the country.

In the selection of an industrial robot, the user has to be concerned about the following:

1. Cost
2. Number and types of axes of motion
3. Power drive
4. Logic
5. Memory
6. Programming
7. Maintenance
8. Environment
9. Physical size and weight
10. Cycle rate

Robot systems charateristics should also be considered and include the following:

1. Manipular work envelope
2. Wrist movement at the end-effector
3. End-effector speed
4. Weight-carrying capacity
5. Type of robot, servo or nonservo
6. Repeatability
7. Power requirements
8. Ambient temperature requirements
9. Interfacing capabilities
10. End-effector tooling
11. Programming support
12. Sensors
13. Physical dimensions and weight

Manufacturers of industrial robots will improve their models. It is the intent of this chapter to present industrial robotic information presently in use.

8.2   Construction and Operation of the Industrial Robot System[a]

A representative industrial robot system is made up of three main parts: a control system, a measuring and servo system, and a mechanical system. The control system consists of the computer, memories, inputs and outputs to control the robot, interlocks from peripherial equipment, and functions to control the servo system of the robot.

The measuring and servo systems include servo amplifiers and dc motors with tachogenerator feedback. Position regulation is by means of a cyclic resolver system, consisting of a resolver with associated supply and decoding circuits, and a position regulator.

The mechanical system includes the robot and the transmission that converts the rotation of the motors into the required motion. The following motions are available:

1. Rotation: (ϕ) entire robot rotates about its pedestal.
2. Out/in motion: (θ) lower arm moves.
3. Up/down motion: (α) upper arm moves.

---

[a] Sections 8.2 through 8.11 are from ASEA, YP 110–302E, YFB September 1976, Req. 6397, Edition 3, courtesy of ASEA, Inc., White Plains, N.Y.

4. Wrist bend: (t) the wrist bends up or down.

5. Wrist turn: (v) the "cuff" of the wrist rotates.

Various additional units (options) can be connected to the control system of the robot:

1. A programming unit, which is used to program robot motions.

2. A tape recorder unit, used to store programs on magnetic tape cassettes.

3. Standby battery supply, ensures that the contents of the program memory are maintained up to 45 min after loss of supply.

4. Memory module to increase the size of the program memory.

5. Test panel for servicing and troubleshooting.

6. Solenoid valves for gripper operation.

7. Gripper.

The control system is based on a computer, as shown in Figure 8.1, which illustrates the main communication paths in the system. The control program is stored in the memory. This program, which is fixed, tells the computer how the various robot instructions and control functions are to be performed. The operator and control system communicate via the control panel and programming unit. The computer regularly scans the state of the controls and reads out acknowledgment orders by switching lamps on or off. The computer monitors and controls input/output signals between the process and the control system, and reads out orders to the axis units if travel is required on any axis.

8.3   Description of Motions

The basic version of the robot has three degrees of freedom for positioning the hand in space. The three main coordinates are a rotary motion (ϕ) and two arm motions in the vertical plane, performed by a lower arm (θ) and an upper arm (α). All three main motions are position-controlled by means of dc motors. A further two degrees of freedom includes a wrist turn and a wrist tilt version.

These motions are available either with dc drive or with pneumatic rotary cylinders with mechanical limit position stops. In the basic version these motions are mechanically blocked.

The drive unit for the rotary motion is rigidly mounted in the pedestal and drives the body around the pedestal column through a reduction gearbox with a very low ratio (see Figure 8.2).

The drive unit for all coordinates except the rotary motion (ϕ) is mounted on the rotating body. The arm motions are provided by ball screws that actuate the arms via links and levers. The lower arm (θ) moves the wrist horizon-

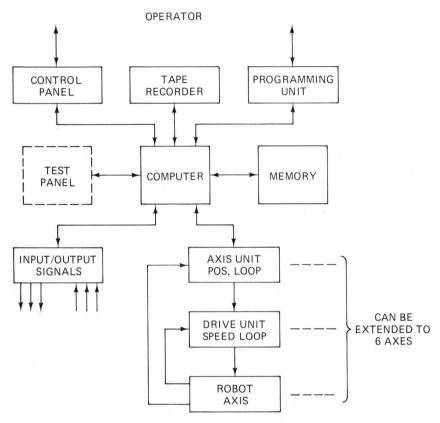

FIGURE 8.1. The operation of the industrial robot system. Courtesy of ASEA, Inc.

tally since the α motion compensates for the θ motion by means of the control system. In the same way, the motion of the upper arm (α) is compensated so that its motion moves the wrist vertically up and down.

The wrist motions are driven via a system of linkages designed so that the wrist coordinates will assume the set angle to the horizontal plane regardless of whether the θ or α coordinate changes. The drive units for the wrist motions are mounted on the frame at the lower bearing point of the lower arm.

The wrist is also designed in such a way that tilting always takes place in the vertical plane.

8.4   Mechanical Linkage and Drives

The drive unit for the rotary motion is rigidly mounted in the pedestal and drives the body around the pedestal column through a reduction gearbox with

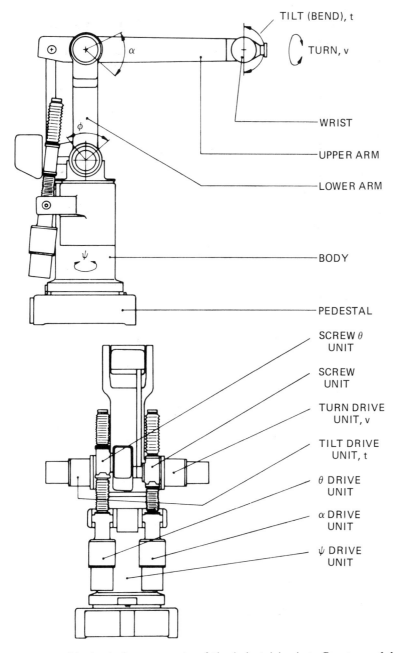

TILT (BEND), t

TURN, v

WRIST

UPPER ARM

LOWER ARM

BODY

PEDESTAL

SCREW θ UNIT

SCREW UNIT

TURN DRIVE UNIT, v

TILT DRIVE UNIT, t

θ DRIVE UNIT

α DRIVE UNIT

ψ DRIVE UNIT

FIGURE 8.2. Mechanical components of the industrial robot. Courtesy of ASEA, Inc.

a very low ratio (Figure 8.3). The tachogenerator is used for velocity information and the resolver for position information.

The drive units for the other coordinates are mounted on the rotating base. The arm motions are provided by ball screws that actuate the arms via links and levers. Figure 8.4 depicts the lower arm and its drive unit. The ball screw moves up and down as the screw turns. The ball screw is connected to a short link that moves the lower arm in and out. In Figure 8.5 the ball screw for the upper arm is shown. Again, as the ball screw moves up and down, it moves the upper arm up and down. The wrist has servo motions, a bend and a tilt.

The body is coupled to the output shaft of the ϕ motion drive unit. It is also mounted on the bearing in the pedestal and rotates on this (the ϕ motion). Mounted on the body are the drive units for lower arm motion (θ motion) and upper arm motion (α).

The electrical wiring from the control cabinet is connected in the pedestal by means of connectors and passes from the pedestal to the body through coiled flexible leads between the pedestal column and the body.

Mounted on the body is a bearing bracket to which the lower arm is fixed and in which it pivots. The motion of the lower arm (θ motion) is obtained by

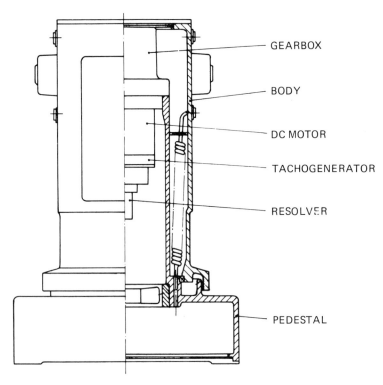

GEARBOX

BODY

DC MOTOR

TACHOGENERATOR

RESOLVER

PEDESTAL

FIGURE 8.3. Pedestal and body. Courtesy of ASEA, Inc.

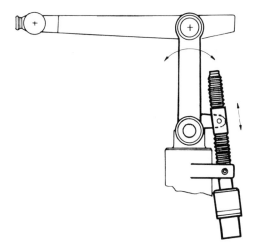

FIGURE 8.4. Lower arm with drive unit. Courtesy of ASEA, Inc.

means of a drive unit consisting of a motor unit and ball screw transmission; these are rigidly mounted on the body. The motion of the ball screw is transmitted to the lower arm via a lever pivoting at the ball screw and attached to the lower arm. The drive units for the turn and tilt motion of the wrist are mounted on the lower arm around its lower turning center.

The upper arm is fixed to, and pivots on, the upper end of the lower arm. Movement of the upper arm is achieved as follows.

The drive unit, consisting of the motor unit and the ball screw transmission, transmits a motion to two link rods articulated on a shaft extending from the ball screw nut assembly. One of the link rods is attached to a shaft fixed to

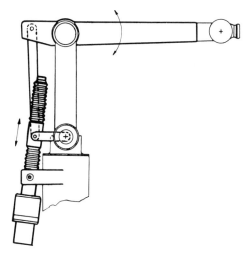

FIGURE 8.5. Upper arm with drive assembly. Courtesy of ASEA, Inc.

the upper arm, and the other end is attached at the center of rotation of the lower arm. Together with the upper and lower arms, these two link rods form a parallelogram. As Figure 8.5 shows, movement of the nut assembly and one corner of the parallelogram will move the wrist up and down.

Electrical and compressed-air lines to control the grippers run through the upper arm. At the rear of the arm there is a connection for the supply of compressed air. Outlets for the gripper, two individually controllable pneumatic outputs, and an electrical contact with four wires are mounted on the underside at the front of the arm.

The wrist linkage system shown in Figure 8.6 operates as follows. A linkage disk is coupled to the drive unit, which is mounted on the lower joint of the lower arm. Mounted on this linkage disk are two link arms displaced $\pi/2$ rad (90°) on the linkage disk. The other end of the link arms is mounted on a linkage disk pivoting in the upper arm. From here the motion is transmitted to a linkage disk in the wrist via two more link rods. The turn and tilt motions have separate linkage systems on either side of the arm system. Wrist tilt is achieved because the linkage disk at the wrist is fixed to the moving part of the wrist. Wrist turn is achieved by means of an angled gear unit that transfers the motion from the linkage disk to a turning disk free to rotate in the wrist.

8.5   Measuring and Servo Systems

Each robot axis is fitted with a resolver for axis position measurement. The resolver has two fixed stator windings and one rotating rotor winding. The resolver stator windings must be supplied with two reference voltages with the same peak-to-peak value and 90° out of phase. The frequency is 2 kHz.

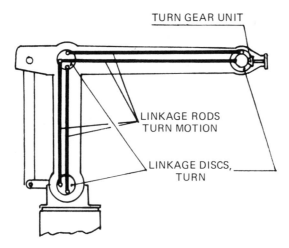

FIGURE 8.6. Schematic diagram of wrist transmission. Courtesy of ASEA, Inc.

The voltage induced in the rotor winding is filtered in the appropriate *axis control circuit* function unit. The phasing of the basic frequency of the resolver response is a measure of the position of the axis. The resolver is coupled to the drive motor shaft. (See Chapter 5 for a detailed description of resolvers.)

A desired value of signal with a frequency of 2 kHz is generated electronically in the axis control circuit function unit. The phase difference between the actual value of signal from the transducer and the desired value of signal is directly proportional to the position error.

During axis motion, the computer regularly reads out orders that result in phase shifting of the desired value of signal. A phase difference between the desired and actual values of signals will cause the robot axis to move and therefore the resolver to rotate in order to reduce the phase difference. During motion at constant speed, the axis lags behind the commanded value with a particular lag that increases with speed. On starting and stopping, the acceleration or retardation is controlled automatically in such a way as to prevent overshoot.

The servo amplifier is part of the position control system. Figure 8.7 shows that a heavy-duty dc voltage is supplied to the drive unit. The positive or negative supply voltage is changed by transistors in the drive unit to a 1-kHz pulsing signal. If the pulsing signal is more positive than negative, the

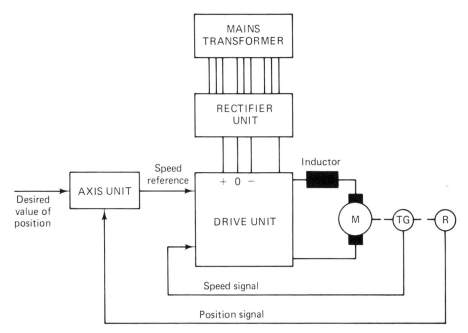

FIGURE 8.7. Block diagram of servo amplifier for one axis [ASEA IRB–6]. Courtesy of ASEA, Inc.

drive motor will turn forward; if more negative, then backward. Figure 8.8 illustrates this concept.

Note that there is an inductor connected in servo with the motor. The purpose of these inductors is to reduce current pulsations due to the pulsating nature of the motor drive current shown in Figure 8.8.

An analog error signal, which serves as the speed reference for the drive motor, is obtained from the axis control circuit function unit. The actual speed is measured by a tachogenerator coupled to the drive motor. The speed reference is compared with the tachogenerator voltage in the speed regulator of the drive unit. The difference is amplified and used as a reference for the motor current. A pulse shaper then controls the transistor switches to give the required direction of motor rotation and the required speed.

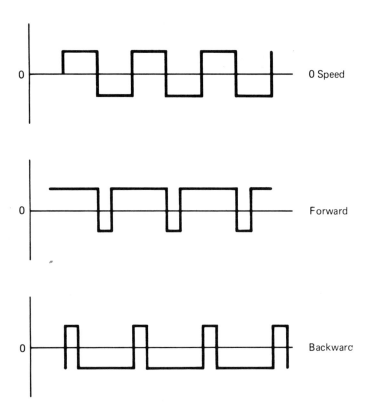

FIGURE 8.8. Pulse signaling nature of servo motor drive current.

8.6   Synchronizing

To enable the patterns of motion of a robot program to be defined relative to the rest of the peripheral equipment, the robot must be set to a well-defined position before the program starts. This takes place on synchronizing to the reference zero point of the robot. When an order for synchronizing is given, the robot axes begin to move at a fixed speed in a positive direction. The axes move until a synchronizing switch (one for each axis) is actuated; when the switch closes, the axis continues to the nearest resolver zero or electrical reference zero point. On synchronizing to a reference zero point, the position registers of the axes are updated so that none of the robot axes can be run up against a mechanical limit position; these position registers are stored in the memory. The working range of the axes is stored in the control program, and if any axis attempts to exceed the value, the motion is interrupted and a warning lamp lights up.

The robot can also be synchronized to its actual position; this means clearing the position error for a brief instant on the axis control circuit function unit. This takes place when the robot is switched on to prevent the robot jerking when the drive motors engage, and on emergency stop for rapid stopping of the robot.

8.7   Electrical Connections

All electrical connections between the control cubicle and the robot run in a flexible steel hose. Signals pass between the control cabinet and the robot for the following:

1. DC motors
2. Resolvers
3. Tachogenerators
4. Synchronizing switches

From three to six electrically controlled axes

5. Solenoid value
6. Limit switches

For pneumatic control of tilt or turn and sixth axis (replaces electrical control

7. Grippers 1 and 2, which are controlled by solenoid valves
8. AC-operated grippers or, alternatively, search devices with an acknowledgment signal
9. Connection leads to connect the programming unit to the robot
10. Protective ground of robot

The programming unit is connected to a socket on the wall of the cabinet.

8.8   Control Structure of an Industrial Robot System

Figure 8.9 is a block diagram of the control equipment. The control equipment consists of a number of function modules built up on circuit boards and in the form of complete units.

The control electronics consist of circuit boards that make up a computer with a memory. There are also interface boards for data communication with the various input and output units (i.e., tape recorder, control panel, programming unit, and axis units). The signals between the function modules of the control electronics appear at the rear plane of the electronics rack and are in the form of a data bus, memory address bus, address pointers for the various input and output units, and a small number of clock pulses and control signals. The digital input and output signals for the control panel, programming unit, and external process go to the input/output unit via the terminal unit. An analog reference signal, the polarity and value of which indicate the direction and speed of movement of the robot axis, is generated in the axis unit. The analog reference signal is amplified in the drive unit so that it can control the dc motor.

This description will now deal first with the computer with memories and interface units. The arrangement of its control program will then be described. Finally, a description will be given of the measuring and servo systems of the robot and the operation of various option modules.

The computer included in the control electronics can be divided up into five main sections:

1. Input for reading in data from the process.
2. Arithmetic unit for arithmetical and logical data processing.
3. Memory to store both data read in from the process and the robot system control program.
4. Control functions to check what is to be executed.
5. Output for the supply of data to the process.

The arithmetic and control sections consist of the function modules known as the central unit and the interruption and power-failure-detection unit. These units contain a clock pulse generator to generate clock pulses to the control section and other electronic sections, a microprocessor, and circuits to control a general 8-bit data bus for data transmission between the microprocessor, memories, and input and output units. The microprocessor contains, among other things, an arithmetical–logical unit, an instruction decoder, and a program counter. The instruction decoder interprets the instruction read from the control program of the memory and establishes the conditions for the execution of the required operation. The program counter

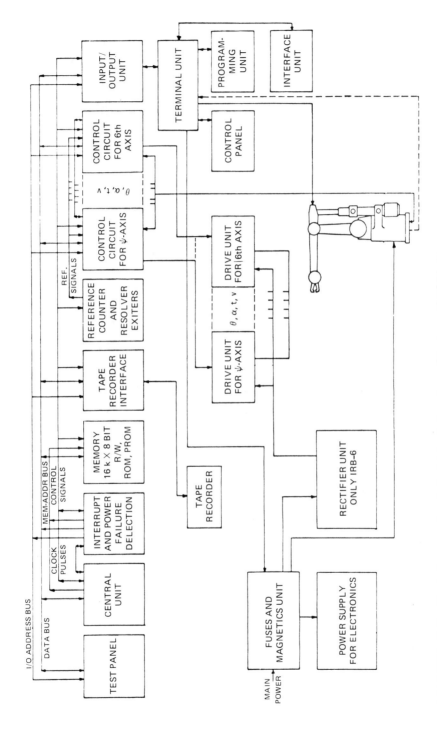

FIGURE 8.9. Block diagram of an industrial robot system. Courtesy of ASEA, Inc.

keeps track of the order in which the computer instructions in the control program are to be carried out.

The memory, which may be either of the write/read type or of the read type only, consists of various combinable function modules, depending on its size and type. The memory is organized in memory words of 8 bits each. A total of 16K (16,384) words can be addressed. There are two electronics modules for the computer memory: the memory unit and the PROM-unit.

The memory unit contains a 4K (4,096) word write/read memory. The write/read memory is a semiconductor memory that stores the robot programs programmed by the operator. It defines the patterns of motion of the robot, that is, data for position coordinates and speeds, and the control of digital inputs and outputs.

The PROM-unit, which contains a 7K (7,168) word read memory, stores the control programs of the systems. The memory content of the PROM-unit does not change in normal service, but it can be reprogrammed with a special programming unit.

Finally, there are the input and output units: these are the tape recorder interfacing, axis control circuits, and input/output unit. The tape recorder interface controls data transmission between magnetic tape and the memory. The axis control circuit function unit receives data words that state by how many increments the relevant robot axis must move. The purpose of the input/output unit is to match words on the bus system of the computer to signals for controlling and checking panel controls and external relay and contact functions. For reference voltage supply of the measuring transducers on the robot, the reference counter and resolver amplifier function unit will be required. A test panel for servicing is also available.

As Figure 8.10 shows, the basic program for controlling the robot system is divided into a number of subprograms. The start-up routine is always activated after switching on; among other things, this clears the registers and program counter. The clock pulse routine synchronizes the computer with the 10-ms sampling period used in the robot system. Whenever the clock pulse arrives, a jump to the clock pulse routine takes place, and at the same time the conditions of a group of controls are read in. While automatic running is in progress, a jump back to the automatic routine always takes place, so new movement data are read out during the next sampling period or at the beginning of execution of the next robot instructions. In other cases, a jump takes place to the mode selection routine. This routine investigates which load is required and whether any related operator actions have been taken. If so, a jump takes place to the routine for the selected mode; if not, the system jumps back to the clock pulse routine to await the next clock pulse.

The manual routine deals with all operator actions relating to the manual mode and required for robot programming, program corrections, synchronizing the robot, and step-by-step running during program testing, for example. When the routine is complete, the system jumps to the clock pulse routine, ex-

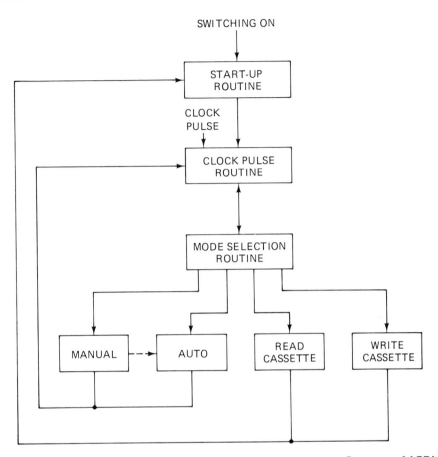

FIGURE 8.10. Flow diagram of basic program of robot system. Courtesy of ASEA, Inc.

cept in step-by-step running, when it jumps to the automatic routine in order to execute the robot instruction.

The automatic routine selects which robot instruction is in turn to be executed. After this, each robot instruction has its own subroutine; in other words, the automatic routine includes a total of 16 subroutines.

The control panel is fitted with controls for switching on the supply voltages, for starting the robot, and for switching between the operational and standby conditions of the robot; it also has controls for normal operation (e.g., program selection and starting or stopping).

The programming unit is used to program the robot. This unit, which is only used at the time of programming, can be connected to several different robot systems.

The programming unit is portable and consists of a box with a programming panel. The programming unit has a 6-m long cable with a plug that can be connected to the programming socket on the control cabinet. The programming unit can be hung up on the control panel of the control cabinet. There is a reel for the cable on the wall of the cabinet.

To avoid tedious reprogramming after long periods of power-supply disconnection, or when more than four programs are to be used with the robot, programs can be stored on cassette tapes. The control equipment must be supplied with an additional electronics unit, the *Tape recorder interface,* and an external plug with the associated wiring to the circuit board.

The standard version of the robot is supplied with electrical wiring to a socket on the robot arm for the connection of search devices or other types of gripper. Different connections must be made in the control cabinet depending on the function for which the socket is to be used. The following signals are available, depending on the application.

Connection of search devices of grippers are:

1. ± 15 V to supply search device electronics (max. 0.5 A).
2. Input to control system for the search–stop function. The search-stop contact or transistor must be dimensioned for + 24 V and be capable of connecting 10 mA to 0 V.

Connection of valves to control grippers includes two make contacts (220 V, 2 A). An optional unit must be added to the control cabinet.

Two solenoid valves to operate two pneumatically controlled grippers are also available as extra equipment. The valves are controlled by signals from the control unit. Compressed-air lines inside the upper arm are fitted to the robot as standard.

8.9   Control Functions

On the control equipment there is a control panel with controls to start the industrial robot system and to select various operating modes. For normal operation of the robot system, that is, when the robot is programmed, only these controls are required. Figure 8.11 shows the controls and indications on the control panel. Some of the principal functions are described in Table 8.1.

8.10   Programming Functions

The portable programming unit, which can be plugged in at the control cabinet or at the robot, is used to reprogram the robot for new applications. When the robot is working in the normal operating mode, the program unit can be used in conjunction with other robot systems. The robot is programmed by using the push buttons on the programming unit to run it to the

FIGURE 8.11. The control panel of the ASEA industrial robot system. Courtesy of ASEA, Inc.

positions it is required to take up in automatic operation. Each position is stored in the program memory when an instruction key on the programming unit is pressed. In addition to positioning instructions, it is possible, for example, to store in the program memory instructions for the operation of grippers, opening and closing of a number of outputs, testing of a number of interlock

TABLE 8.1
SWITCH AND FUNCTION

Switch	Function
Main switch	Control main power supply
Program selection	Can select one of four stored programs
Synchronizing	Robot axes are sent to their synchronizing position
Emergency stop	Stops all robot motions
Operating mode	Can be either MANUAL mode, where programming box controls robot, or AUTO mode, where program executes
Program start	If in AUTO mode, program will start to execute

[Courtesy of ASEA, Inc.]

inputs, time lag, and repeating. The programmed instructions are then executed in sequence in automatic mode. Some of the programming unit is shown in Figure 8.12. Some of the principal instructions and controls are given in Table 8.2.

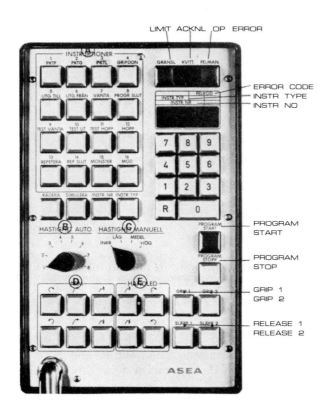

A INSTRUCTIONS	10 TEST OUT	B AUTO SPEED
	11 TEST JUMP	
	12 JUMP	C MANUAL SPEED
1 PTP F	13 REPEAT	[Incr; Low;
2 PTP C	14 END REPEAT	Medium; High]
3 PTP L	15 PATTERN	
4 GRIPPER	16 MOD	D ARM
5 OUTPUT ON		
6 OUTPUT OFF	DELETE	E WRIST
7 WAIT	SIMULATE	
8 PROGRAM END	INSTR NO	
9 TEST WAIT	INSTR TYPE	

Note: With switch A in the circuit, the top system can be used. With switch B in the circuit, the bottom system can be used.

FIGURE 8.12. Programming unit for the ASEA industrial robot system. Courtesy of ASEA, Inc.

Push Button	Function
INSTRUCTIONS	INSTRUCTIONS is a group of push buttons used to program the instructions to be executed when the robot is operating automatically. On programming, each instruction in the program is given a serial number. The instructions themselves are designated by a type number from 1 to 16.
[a] PTP F	Point-to-point operation with maximum accuracy. With arguments [special numerical values], search and vertical/horizontal running can be programmed.
[b] PTP C	Point-to-point operation with course positioning. With arguments [special numerical values], search and vertical/horizontal running can be programmed.
[c] PTP L	Straight-line operation such that positioning is concluded simultaneously for all axes. The travel time is stated in the form of an argument.
[d] GRIPPERS	The states of the grippers [i.e. "on" or "off"] are stored.
[e] OUTPUT ON	The digital output [1–14] selected on the keyboard must be "on."
[f] OUTPUT OFF	The digital output [1–14] selected on the keyboard must be "off."
[g] WAIT	Waiting time as set on the keyboard [0.1–9.9 s].
[h] PROGRAM END	End of robot program. Restart from first instruction or jump to next program.
[i] GRIP 1	Gripper 1 closes.
[j] RELEASE 1	Gripper 1 opens.
[k] ARM	Used for manual operation of the ϕ, θ, and α axes of the robot. The travel speed is selected with the MANUAL SPEED switch. Several axes can be operated simultaneously. When only one of α and θ are operated, the control system compensates to give a purely vertical or horizontal motion.
	ϕ axis rotates clockwise
	ϕ axis rotates counterclockwise
	θ axis moves forward
	θ axis moves back
	α axis moves up
	α axis moves down
[l] WRIST	Used for manual operation of the robot wrist. Function as described for the ARM pushbuttons.
	The tilt motion is compensated with the turn motion to give pure bending.
	tilt axis moves up
	tilt axis moves down
	turn axis rotates clockwise
	turn axis rotates anticlockwise
[m] PROGRAM START	Used in automatic operation to start the robot program.
	In manual operation, new programs can be tested step by step, since one instruction is performed every time the push button is pressed.
[n] PROGRAM STOP	Used to stop programs or instructions in progress.
Digital Keyboard	Function
[a] 0, 1, 2, 3, 4, 5, 6, 7, 8, 9	Used to key in numerical values and arguments belonging to the relevant instruction and to select the instruction for indication or deletion.
[b] R	Clearing before input; the numerical display panel is also cleared.

FIGURE 8.12 [continued].

TABLE 8.2
SWITCH, PUSHBUTTON, AND DIGITAL KEYBOARD AND FUNCTIONS

Switch	Function
Auto speed	Selection of speed to be programmed for positioning in automatic mode. In automatic running of program, the highest speed is limited to the speed to which this switch is set. In step-by-step operation, the speed is limited to not more than speed 6. Position: 8, maximum speed 7, 70% of max. speed 6, 50% of max. speed 5, 31% of max. speed 4, 13% of max. speed 3, 5% of max. speed 2, 2.5% of max. speed 1, 1.3% of max. speed
Manual speed	Speed selection for manual operation of the robot. The motion is executed by means of the ARM and WRIST push buttons.
[a] INCR	The relevant robot axis moves one increment.
[b] LOW	The relevant robot axis moves at 1.3% of maximum speed.
[c] MEDIUM	The relevant robot axis moves at 15% of maximum speed.
[d] HIGH	The relevant robot axis moves at 50% of maximum speed.

[Courtesy of ASEA, Inc.]

8.11   Specifications for the ASEA IRB–6 and IRB–60 Industrial Robot System

Technical data for the ASEA IRB–6 and IRB–60 follow:

Robot Characteristics	IRB–6	IRB–60
1. Permitted handling weight including gripper:	6 kg	60 kg
2. Maximum gripper length at above handling weight:	200 mm	400 mm
3. To calculate the permissible handling weight when the gripper length exceeds 200 mm or 400 mm as the case may be, the following mechanical data apply:		
Maximum moment of inertia	2.5 Nm^2	100 Nm^2
Maximum static load	12 Nm	240 Nm

Robot Characteristics	IRB-6	IRB-60
4. The robot is available with the following degrees of freedom:		
Arm motion: rotation	340°	330°
Arm motion: radial	±40°	±50° to −20°
Arm motion: vertical	±25° to −40°	+10° to −55°
Wrist motion: bend	±90°	+75° to −120°
Wrist motion: turn	±180°	±180°

5. Gripper functions: Two independent solenoid valves housed in the upper arm, which can be operated from the keyboard of the programming unit, are available as an option. There is also a four-pole electrical outlet in the upper arm; among the intended applications of this is the supply of more advanced grippers with search functions.

6. Weight of robot:	125 kg	750 kg
7. Repeat accuracy at wrist:	±0.20 mm	±0.40 mm
8. Speeds:		
Arm motion: rotation	95°/s	90°/s
Arm motion: radial	0.75 m/s	1.0 m/s
Arm motion: vertical	1.1 m/s	1.35 m/s
Wrist motion: bend	115°/s	90°/s
Wrist motion: turn	195°/s	150°/s
9. Number of servo-driven degrees of freedom:		
	5	5

10. Environment:
    The robot and the controls mounted on it are designed to withstand ambient temperatures from +5° to +70°C. The rating of the dc motors applies up to ambient temperatures of 50°C. For ambient temperatures above 50°C, the permitted handling capacity must be determined in consultation with ASEA in each individual case.

11. Dimensions and working space are given in Figure 8.13 through 8.16.

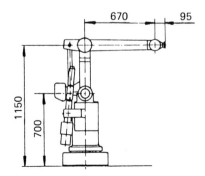

FIGURE 8.13. Dimension drawing of industrial robot with 6-kg handling capacity.

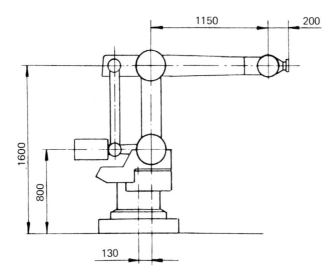

FIGURE 8.14. Dimension drawing of industrial robot with 60-kg handling capacity.

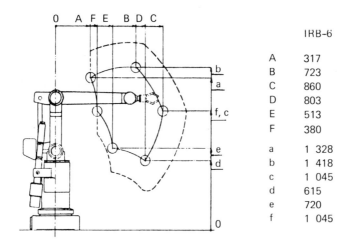

	IRB-6
A	317
B	723
C	860
D	803
E	513
F	380
a	1 328
b	1 418
c	1 045
d	615
e	720
f	1 045

FIGURE 8.15. Working range of industrial robot [IRB-6]. Courtesy of ASEA, Inc.

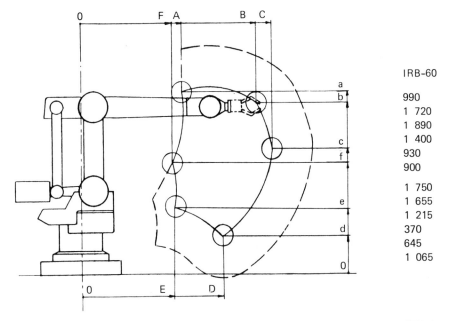

IRB-60

a	
b	990
	1 720
	1 890
	1 400
c	
f	930
	900
	1 750
e	1 655
	1 215
d	370
	645
	1 065
0	

FIGURE 8.16. Working range of industrial robot (IRB-60). Courtesy of ASEA, Inc.

12. Power drives for the ASEA IRB-6 and IRB-60: electric.

13. *Control system for robot with 6-kg handling capacity*

Cabinet dimension [width × depth × height]	720 × 720 × 1620 mm
Weight	IRB-6: 325 kg IRB-60: 425 kg
Permitted ambient temperature	0–40°C
Permitted relative humidity	5–90%
Mains supply voltage for control equipment and robot	380 V, 415 V (425 V) or 440 V (3-phase)
Total power consumption including robot	IRB-6: 0.8–1.7 kW IRB-60: 0.8–7 kW
Mains frequency	50 or 60 Hz
Permitted mains voltage variation	±10%
Permitted frequency variation	±1 Hz
Number of signal outputs	14
Number of interlock inputs	16
Type of position regulation	Closed with resolver as measuring transducer
Number of electrically controlled axes, max.	6
Number of programs in program memory	4

Programming capacity of basic version	At least 250 instructions
Standby battery supply of program memory for mains voltage failure	At least 45 min
Maximum distance between control cabinet and robot	15 m

14. *Programming units*

Dimension [width × height × depth]	182 × 314 × 69 mm
Permitted ambient temperature	0–45°C
Permitted relative humidity	5–90%
Weight	4 kg

Portable and with 6-m cable for connection to control cabinet or robot

15. *Tape unit*

Dimensions [width × height × depth]	153 × 165 × 394 mm
Weight	6.5 kg
Permitted ambient temperature	5–35°C
Permitted relative humidity	20–90%

Portable with 2-m cable; can be connected to control cabinet

Technical data for the IRB–60/2 follow:

1. Number of servo-driven degrees of freedom   5
2. Performance

Type of movement	Working range	Max. speed
Rotational movement C	330°	90°/s
Radial arm movement B		1.0 m/s
Vertical arm movement A		1.35 m/s
Wrist bending movement E	+75 to −120°	90°/s
Wrist rotational movement P	+/−180°	150°/s
Handling capacity including gripper	Max. 60 kg	
Incremental movement	Approx. 0.2 mm	
Repetition accuracy at wrist center	≤ ±0.4 mm	
Max. moment of inertia	10 kg^2	
Max. static load	240 Nm	
Power consumption		
Operation	0.8–3.6 kW	
Standby	0.5 kW	

3. Environmental Requirements

Protection standard	IP 54
Ambient temperature	
Control equipment, operation	+5 to +45°C
Mechanical robot and motors	+5 to +50°C
Upper arm and wrist	+5 to +80°C
Relative humidity	Max. 90%

4. Physical Data

Weight	
Mechanical robot	885 kg
Control cabinet	350 kg
Volume	
Mechanical robot, H × W × D	1860 × 1265 × 1650 mm
Control cabinet, H × W × D	1900 × 820 × 700 mm

5. Electrical Connections

Power supply	
Mains voltage	3-phase, 380 V, 10 ± 15%
Frequency	50 Hz ± 1 Hz
Supply voltage available for	Nom. 24 V dc,
optional connections	max. 2 A
Digital connections	
Inputs	7; rated voltage 24 V dc; impedance 3.5 k Ω
Outputs	6 (+2 for gripper device). Current-feeding rated Voltage 24 V dc Load 150 mA
Signal connections	
Input signals	Opening contact to give emergency stop. Supply 24 V
Output signals	Contact open when emergency stop activated. Load 1 A/60 V
	Contact closed when front door is open. Supply 220 V/270 mA
	Contact closed in operation mode. Supply 220 V/135 mA

6. Program Capacity
   Number of programs in user memory
       Main program   1
       Subprograms ≤ 999
   Memory
       Capacity   6 Kword
       Number of positioning instruc-   Approx. 470
           tions only (if external axes
           included, number of instruc-
           tions is reduced by 40%)
   Battery backup   Approx. 100 h
   Recharging time   Approx. 24 h
   External memory backup   Nom. 24 V dc, 10 mA
   Work point (TCP)
       Number definable   9
   Position register
       Number definable   100
   Register
       Number definable   100
       (may contain four-digit
       numbers with sign)
   Sensors, possible connections
       (Requires adaptive control pro-
       gram, see variants of basic sys-
       tem.)
           Number addressable   16
       Sensors may be of digital type,
       with up to 8 bits + 1 sign bit, or
       of analog type. Connected to
       digital or analog inputs in
       system.

8.12   Recent Robots Manufactured by ASEA, Inc.[b]

The industrial robot system of the IRB–60/2 consists of the basic system and accessories shown in Figure 8.17. The basic model, which is the mechanical robot (Figure 8.18), together with a control cabinet can be supplied in a number of variants.

---

[b] The material in Sections 8.12 and 8.13 is used courtesy of ASEA, Inc., White Plains, N.Y.

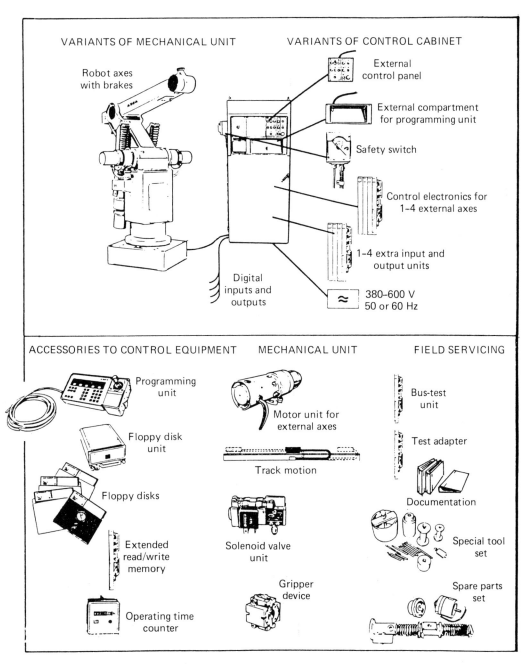

FIGURE 8.17. Basic robot system of the IRB–60/2. Courtesy of ASEA, Inc.

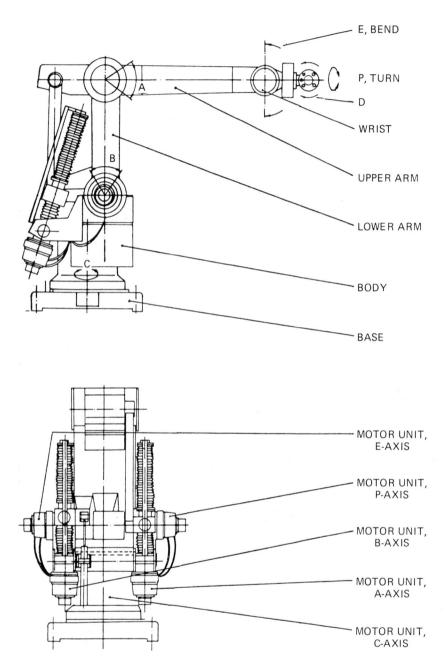

E, BEND

P, TURN

D

WRIST

UPPER ARM

LOWER ARM

BODY

BASE

MOTOR UNIT,
E-AXIS

MOTOR UNIT,
P-AXIS

MOTOR UNIT,
B-AXIS

MOTOR UNIT,
A-AXIS

MOTOR UNIT,
C-AXIS

FIGURE 8.18. IRB–60/2 mechanical robot. Courtesy of ASEA, Inc.

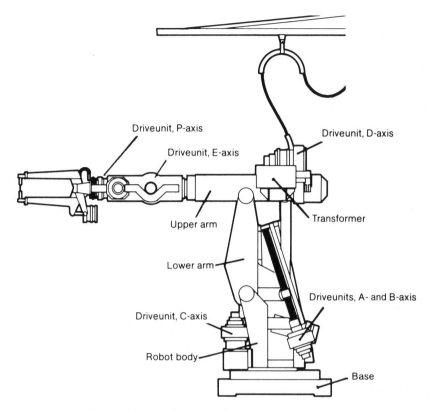

Driveunit, P-axis

Driveunit, D-axis

Driveunit, E-axis

Upper arm

Transformer

Lower arm

Driveunits, A- and B-axis

Driveunit, C-axis

Robot body

Base

FIGURE 8.19. The IRB 90/2. Courtesy of ASEA, Inc.

ASEA's new robot system, the IRB–90/2, shown in Figure 8.19, for spot welding is designed to fulfill all the requirements made of a flexible spot welding set. It is easy to install, and it will reduce running costs and provide a high degree of utilization of capital. The system consists of robot (mechanical part), control equipment, welding timer, welding gun, mast unit, cables, and hoses.

8.13 Applications of the ASEA Industrial Robots

Applications of the ASEA industrial robots include the following:

1. Deburring
2. Polishing

3. Cutting
4. Investment casting
5. Inspection
6. Arc welding
7. Spot welding
8. Forging
9. Machine tool loading and unloading
10. Parts transfer
11. Machining
12. Gluing
13. Snag grinding

Photos of the ASEA IRB–6 and IRB–60 are showbn in Figures 8.20 and 8.21.

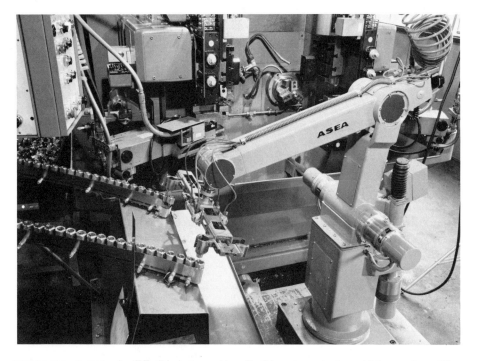

FIGURE 8.20. An IRB–6 robot used in a flexible automated machining system. The robot loads and unloads all machines and inspection fixtures in the system. It is equipped with a double gripper. This enables it to unload a part from a machine with an empty gripper and load a part into the machine from the other gripper. Courtesy of Industrial Robot Division, ASEA, Inc.

FIGURE 8.21. The IRB–60 arc welding a portion on a truck cab assembly. The robot can manipulate the arc welding gun into areas with limited access, as seen here. Total arc welding cycle time is about one-half that required when arc welding is done manually. Courtesy of Industrial Robot Division, ASEA, Inc.

8.14   Specifications and Applications for Unimation Robots[c]

The 260 Series PUMA has six axes of motion, a repeatability of 0.05 mm, and is recommended for medium- to high-speed assembly and material handling applications where tight tolerance is required. It is electrically power driven.

Load capacity of the pneumatically operated gripper is 5.0 lb (2.3 kg), which includes the gripper. Arm tip velocity is 3.3 ft/s (1.0 m/s) with maximum load. The 260 Series can be taught a program with a standard teach pendant or VAL Computer Control, the proprietary language for Unimate and PUMA robots designed and developed by Unimation, Inc.

Unimation, Inc., has also available the all-electric PUMA 560 Series robot with the speed, repeatability, and reach for a wide range of assembly, finishing, handling, joining, and inspection operations.

[c] The material in Section 8.14 is used courtesy of Unimation[®], Inc., Danbury, Conn.

Equipped with sophisticated VAL Computer Control, the PUMA 560 is typically used in packaging, palletizing and machine-loading applications requiring its positional accuracy and repeatability.

A six-axis robot, the 560 has a 40-in. (1.0-m) reach and a repeatability of 0.004 in. With the dexterity to match its other capabilities, the robot is ideally suited to numerous applications involving quality control and inspection where it can be easily integrated into existing work stations. See Figure 8.22 for the PUMA 260 Series, 550 Series, and 560 Series robots. The PUMA 260 Series, 550 Series, and 560 Series PUMA arm specifications are shown in Figure 8.23.

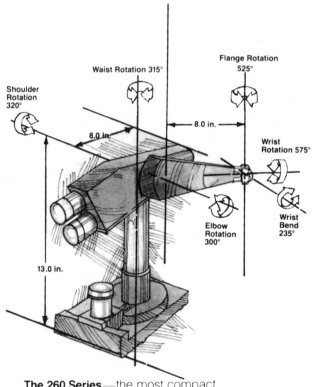

The 260 Series—the most compact PUMA robot, offers the greatest speed and repeatability. With six axes of motion and the powerful computer control, the 260 Series is recommended for medium to high speed assembly and material handling applications.

FIGURE 8.22. The PUMA™ 260 series, 550 series, and 560 series robots. Courtesy of Unimation®, Inc., Danbury, Conn.

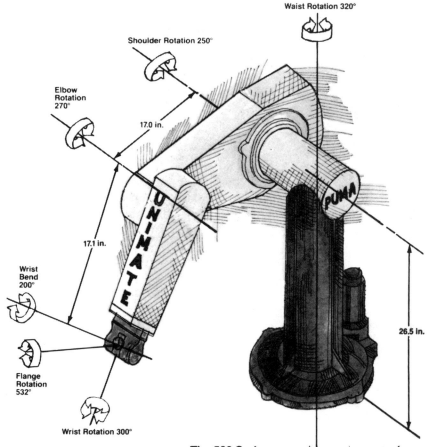

Waist Rotation 320°

Shoulder Rotation 250°

Elbow Rotation 270°

17.0 in.

17.1 in.

Wrist Bend 200°

Flange Rotation 532°

Wrist Rotation 300°

26.5 in.

The 560 Series—combines six axes of motion with 40 inch reach to easily perform assembly, finishing, handling, joining and inspection operations.

FIGURE 8.22. [continued].

The new all-electric PUMA 760 Series robot is capable of increased payload capacity and extended reach. The latest advance in the sophisticated PUMA robot line, the 760 incorporates VAL Computer Control, and advanced, plain English programming language and control system. The 760 was designed and developed by Unimation to meet industry demand for a medium-sized, electric, computer-controlled robot for applications requiring dexterity and high accuracy.

A six-axis robot, it has a 49-in. (1.25-m) reach, a 22-lb (10-kg) payload capacity, and repeatability to ± 0.008 in. (0.2 mm). The PUMA 760 can be applied to a wide range of assembly, joining, inspection, finishing, and material-

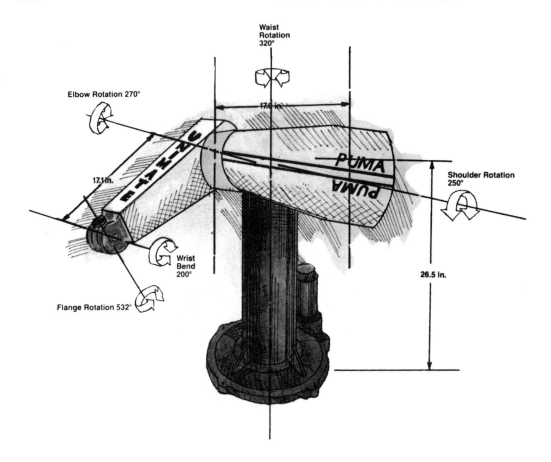

The 550 Series—features five axes of motion, a reach of nearly 40 inches and a load capacity of approximately five pounds. It's capable of handling most industrial assembly, transfer or packaging operations.

FIGURE 8.22. [continued].

handling applications to increase productivity and lower manufacturing costs. Specifications for the PUMA 760 Series are shown in Figure 8.24.

The Unimate 1000, 2100B, 2000C, 2100C, and 4000B specifications are shown in Figure 8.25.

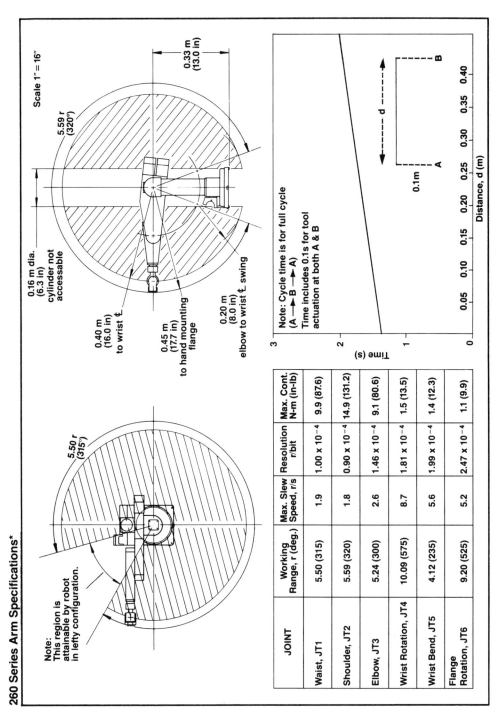

260 Series Arm Specifications*

Scale 1″ = 16″

5.59 r (320°)

0.33 m (13.0 in)

0.16 m dia. (6.3 in) cylinder not accessible

0.40 m (16.0 in) to wrist ₵

0.45 m (17.7 in) to hand mounting flange

0.20 m (8.0 in) elbow to wrist ₵ swing

5.50 r (315°)

Note: This region is attainable by robot in lefty configuration.

Note: Cycle time is for full cycle (A → B → A) Time includes 0.1s for tool actuation at both A & B

d

0.1m

A   B

Distance, d (m)

Time (s)

JOINT	Working Range, r (deg.)	Max. Slew Speed, r/s	Resolution r/bit	Max. Cont. N-m (in-lb)
Waist, JT1	5.50 (315)	1.9	1.00×10^{-4}	9.9 (87.6)
Shoulder, JT2	5.59 (320)	1.8	0.90×10^{-4}	14.9 (131.2)
Elbow, JT3	5.24 (300)	2.6	1.46×10^{-4}	9.1 (80.6)
Wrist Rotation, JT4	10.09 (575)	8.7	1.81×10^{-4}	1.5 (13.5)
Wrist Bend, JT5	4.12 (235)	5.6	1.99×10^{-4}	1.4 (12.3)
Flange Rotation, JT6	9.20 (525)	5.2	2.47×10^{-4}	1.1 (9.9)

FIGURE 8.23. The PUMA™ 360 series, 550 series, and 560 series robot arm specifications. Courtesy of Unimation®, Inc., Danbury, Conn.

550/560 Series Arm Specifications*

Note:
This region is attainable by robot in lefty configuration.

320°

Scale 1″ = 38″

0.15 m dia. (5.9 in) cylinder not accessible

4.72 r (250°)

0.67 m (26.5 in)

0.86m (34.1) to wrist ₵

0.92 m (36.3 in) to hand mounting flange

0.43 m radius (17.0 in.) elbow to wrist ₵ swing

Note: Cycle time is for full cycle (A → B → A)
Time includes 0.1s for tool actuation at both A & B

0.1m

A

d

B

Time (s)

Distance, d (m)

JOINT	Working Range, r (deg.)	Max. Slew Speed, r/s	Resolution r/bit	Max. Stall, N-m (in-lb)
Waist, JT1	5.59 (320)	1.4	1.00×10^{-4}	67 (590)
Shoulder, JT2	4.72 (250)	0.9	0.73×10^{-4}	113 (1000)
Elbow, JT3	4.72 (270)	2.1	1.17×10^{-4}	57 (500)
Wrist Rotation, JT4 (Optional)	4.89 (280)	4.0	0.83×10^{-4}	14 (120)
Wrist Bend, JT5	3.49 (200)	2.1	0.87×10^{-4}	12 (110)
Flange Rotation, JT6	9.29 (532)	7.9	0.82×10^{-4}	14 (120)

*All specifications subject to change without notice.

PUMA 760 Series Specifications

General

Configuration: 6 Revolute axes
Drive: Electric DC Servos
Controller: System computer (LSI-11/2)
Teaching Method: By manual control and/or computer terminal.

Program
   Language: VAL
Program Capacity: 16K RAM user memory std.
External Program
   Storage: Floppy-disk (optional)
Gripper Control: 4-Way pneumatic solenoid
Power
   Requirement: 220 or 440 volt, 50-60 Hz, 6300 watts

Optional
   Accessories: CRT or TTY terminals, floppy disk memory storage, CMOS Memory, I/O module (8 input/8 output signals isolated AC/DC levels) up to 32 I/O capability (4 modules)

Performance

Repeatability: ± 0.2mm (0.008 in)
Load Capacity: *At flange rotation*, 14.1 in-oz-sec^2
At wrist bend, 56.7 in-oz-sec^2

Straight Line
   Velocity: 1.0 m/s (40 in/s) max.

Environmental Operating Range

10–50°C (50–120°F)
10-80% relative humidity (non-condensing).
Shielded against industrial line fluctuations and human electrostatic discharge.

Physical Characteristics

Arm Weight: 2225N (500 lbs.)
Controller Size: 1.10 x 1.04 x 0.66m
Controller Weight: 1780N (400 lbs.)
Controller Cable
   Length: 15.24m (50ft.) max.

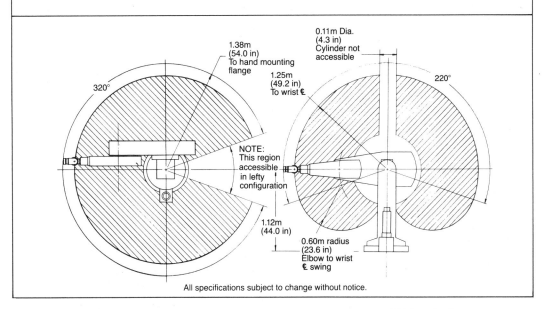

All specifications subject to change without notice.

FIGURE 8.24. The PUMA™ 760 series specifications. Courtesy of Unimation®, Inc., Danbury, Conn.

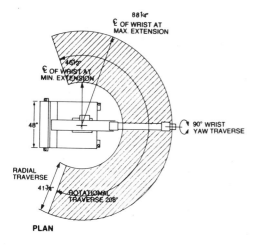

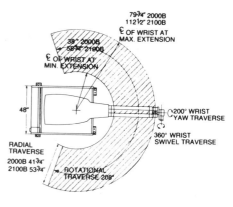

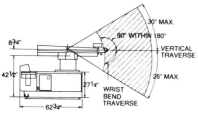

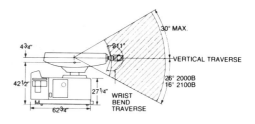

1000

2000B-2100B
COMPOSITE DRAWING

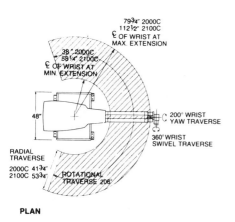

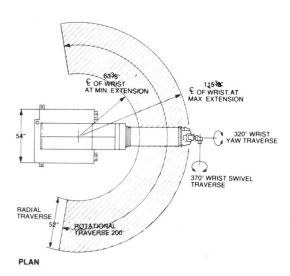

FIGURE 8.25. Unimate 1000, 2000B, 2100B, 2000C, 2100C, and 4000B specifications. Courtesy of Unimation®, Inc., Danbury, Conn.

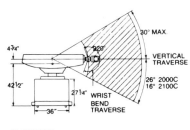

ELEVATION

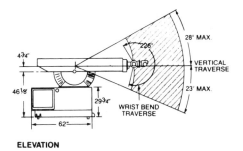

ELEVATION

2000C-2100C
COMPOSITE DRAWING
CONTROL CONSOLE 30"H x 48½"W x 18"D
HYDRAULIC SUPPLY 29"H x 48½"W x 28½"D

4000B
CONTROL CONSOLE 30"H x 48½"W x 19"D

	1000	2000B	2100B	2000C	2100C	4000B
Mounting Position	Floor	Floor	Floor	Any	Any	Floor
No. of Degrees of Freedom	3 to 5	3 to 6	3 to 6	3 to 6	3 to 6	3 to 6
Positioning Repeatability	0.05 in (1.27 mm)	0.05 in (1.27 mm)	0.08 in (2.03 mm)	0.05 in (1.27 mm)	0.08 in (2.03 mm)	0.08 in (2.03 mm)
Power Requirements	460V,3φ 60Hz,9kVA	460V,3φ 60Hz,11kVA	460V,3φ 60Hz,11kVA	460V,3φ 60Hz,11kVA	460V,3φ 60Hz,11kVA	460V,3φ 60Hz,34kVA
Maximum Lift	50 lb (22.9 kg)	300 lb (136 kg)	300 lb (136 kg)	300 lb (136 kg)	270 lb (123 kg)	450 lb (205 kg)
Standard Wrist Torque						
Bend	500 in-lb (5.7 kg-m)	1000 in-lb (11.5 kg-m)	1000 in-lb (11.5 kg-m)	1000 in-lb (11.5 kg-m)	1000 in-lb (11.5 kg-m)	3500 in-lb (40.3 kg-m)
Yaw	150 in-lb (1.7 kg-m)	600 in-lb (6.9 kg-m)	600 in-lb (6.9 kg-m)	600 in-lb (6.9 kg-m)	600 in-lb (6.9 kg-m)	2800 in-lb (32.2 kg-m)
Swivel	N/A	800 in-lb (9.2 kg-m)	800 in-lb (9.2 kg-m)	800 in-lb (9.2 kg-m)	800 in-lb (9.2 kg-m)	2300 in-lb (26.5 kg-m)
Heavy Duty Wrist Torque						
Bend	N/A	2000 in-lb (23 kg-m)	2000 in-lb (23 kg-m)	2000 in-lb (23 kg-m)	2000 in-lb (23 kg-m)	11000 in-lb (126.5 kg-m)
Yaw	N/A	1200 in-lb (13.8 kg-m)	1200 in-lb (13.8 kg-m)	1200 in-lb (13.8 kg-m)	1200 in-lb (13.8 kg-m)	2800 in-lb (32.2 kg-m)
Swivel	N/A	800 in-lb (9.2 kg-m)	800 in-lb (9.2 kg-m)	800 in-lb (9.2 kg-m)	800 in-lb (9.2 kg-m)	N/A
Memory Options						
Point to Point	Up to 256 points	Up to 2048 points	Up to 2048 points	Up to 2048 points	Up to 2048 points	Up to 2048 points
Continuous Path	N/A	Up to 500 in of travel	Up to 500 in of travel	Up to 500 in of travel	N/A	N/A
Environment	40°F (5°C) to 120°F (50°C), Humidity 0–90%; all models.					

8.15 Specifications and Applications for Cincinnati Milacron Robots[d]

The T^3–566 or T^3–586 robots are electrohydraulic power driven. Milacron robots that are electrically driven are the T^3–726, T^3–746, and T^3–756 series. The specifications for the T^3–566 or T^3–586 are shown in Figure 8.26. Applications for Milacron robots are shown in Figure 8.27 through 8.30. The T^3–566 or T–586 Cincinnati Milacron robots are used in forging, investment casting, machine tool loading and unloading, parts transfer, finishing, plastic molding, welding, machining, and inspection.

The electric-driven computer-controlled T^3–746 industrial robot is designed to boost productivity and is used in arc welding, inspection, drilling, routing, deburring, grinding, material handling, polishing, and gluing.

The Milacron robot arc welding system with Milacron Seam Tracking provides total interfacing among the elements of the system. Power-supply wire feeder, positioning tables, and associated hardware are all interfaced with the robot arm through the computer control.

The Weave Function provides the operator with a selection of programmable weave patterns while eliminating the need for mechanical oscillators following the seam. Align to seam capability enables the operator to easily shift the weld gun from the programmed start point to the actual start point. The robot control automatically adjusts the entire program in accordance with the shift.

8.16 Review Questions

1. Discuss the mechanical linkage and drives of an industrial robot system.
2. Discuss the electrical connections of an industrial robot system.
3. Discuss the programming functions of an industrial robot system.
4. Discuss the specifications of two industrial robots used with electrical and electrohydraulic power drive systems.
5. Discuss the applications of two industrial robots used with electrical and electrohydraulic power drive systems.

---

[d] The material in Section 8.15 is used courtesy of Cincinnati Milacron Industrial Robot Division, Lebanon, Ohio.

Load capacity
Load 10 in (250 mm) from tool mounting plate, T^3-566 Robot ..... 100 lb (45 kg)[1]
Load 10 in (250 mm) from tool mounting plate, T^3-586 Robot .... 225 lb (100 kg)[1]

Number of axes, control type
Number of servoed axes, hydraulically driven 6
Control type .... Controlled path at tool center point

Positioning repeatability
Repeatability to any programmed point .... ±0.050 in (±1.25 mm)[2]

Jointed arm motions, range velocity
Maximum horizontal sweep .... 240°
Maximum horizontal reach to tool mounting plate,
  T^3-566 Robot .... 97 in (2460 mm)
Maximum horizontal reach to tool mounting plate,
  T^3-586 Robot .... 102 in (2590 mm)
Minimum to maximum vertical reach .... 0 to 154 in (0 to 3900 mm)
Maximum working volume .... 1000 cu ft (28 m^3)
Maximum velocity of tool center point (TCP), T^3-566 Robot .. 50 ips (1270 mm/s)
Maximum velocity of tool center point (TCP), T^3-586 Robot ... 35 ips (890 mm/s)
Pitch .... 180°
Yaw .... 180°
Roll .... 270°

Memory capacity
Number of points which may be stored 1,750

Floor space and approximate net weight
Robot .... 9 sq ft (0.8 m^2); 5,000 lb (2,270 kg)
Hydraulic and electric power unit .... 17 sq ft (1.5 m^2); 1,900 lb (540 kg)
ACRAMATIC control .... 8.3 sq ft (0.8 m^2); 800 lb (365 kg)
Power requirements .... 460 volts, 3 phase, 60 Hz[3]
Environmental temperature .... 40 to 120°F (5 to 50°C)[1]

[1]Consult factory for special applications.
[2]T^3-566 Robot available with ±0.025 in (±.65 mm) repeatability at 15 ips) 375 mm/sec).
[3]50 Hz available.

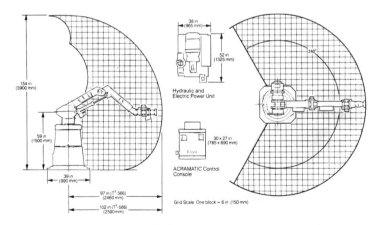

FIGURE 8.26. The Cincinnati Milacron T^3-566 or T^3-586 robot specifications. Courtesy of Cincinnati Milacron Industrial Robot Division, Lebanon, Ohio.

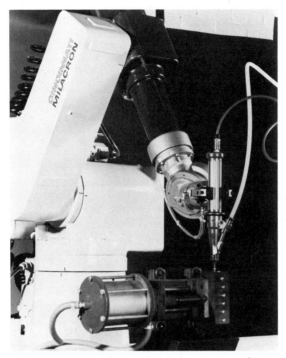

FIGURE 8.27. Here Cincinnati Milacron's new electric T³-726 Industrial Robot uses an automatic screwdriver to insert screws into a workpiece. The ability of the robot to maintain correct tool orientation is essential in this application. Courtesy of Cincinnati Milacron Industrial Robot Division, Lebanon, Ohio.

FIGURE 8.28. In this production cell at International Harvester's Farmall Division, a Cincinnati Milacron computer-controlled T³ Industrial Robot serves two Milacron Cinturn Series C turning centers. Parts enter the cell as rough casting and leave it completely machined and inspected.

FIGURE 8.29. In an appliance manufacturing plant, this Cincinnati Milacron computer-controlled T³ Industrial Robot removes refrigerator liners from a trim press and hangs them on hooks on a moving conveyor. The robot uses Milacron's tracking option to follow the conveyer, maintaining continuous knowledge of where the hooks are located, and thus maintaining the correct relationship between each refrigerator liner and hook position. Courtesy of Cincinnati Milacron Industrial Robot Division, Lebanon, Ohio.

FIGURE 8.30. Here, at a General Motors Assembly Division plant in South Gate, California, fourteen Cincinnati Milacron computer-controlled T³ Industrial Robots spot weld J-car bodies at a rate of 72 bodies per hour. The robots spot an average of 400 welds on each body. Four different models of J-car bodies are spot welded on the same line. Courtesy of Cincinnati Milacron, Lebanon, Ohio.

8.17   Bibliography

The 1983 Robotics Industry Directory™, Technical Database Corporation, Conroe, Texas; published yearly.

Robotics Today '82, Annual Edition, 368 pages, Society of Manufacturing Engineers, 1982.

Edward L. Safford, Jr., *The Complete Handbook of Robotics,* Tab Publishers, Summit, Pa.

Robot Software Systems

9.1   Introduction

Computer hardware, to be useful and effective, must be supported by extensive software. In fact, dozens of computer languages are available for work in various application areas. FORTRAN, COBOL, BASIC, and Pascal are a few of the popular general-purpose languages, while ECAP, NASTRAN, and STRESS are a sampling of the popular special-purpose languages. The same situation (i.e., a multitude of languages) exists for robots.

At the present time, there are more than 14 programming languages used by industrial robots. Table 9.1 gives the robot language, manufacturing listing, and robot arm listing used today. More languages and robot software systems will be necessary tomorrow. In the future, small educational robots may be required to mimic the industrial robot. On the other hand, software systems for industrial robots are complex. Such software programs are useful in robot vision systems and robot work cells.

It may be possible to program a small robot such as the Minimover 5 of Microbot, Inc., to mimic the control language for a PUMA–500 unit. Programs on the Minimover can be transferred to the PUMA controller for execution.

The writing of a program for a robot includes the following:

1. Definition of problem to be solved.
2. Best technique of the solution.
3. Flow chart needed for the detailed analysis of the solution.
4. Computer programming language coding for the solution of the problem.

TABLE 9.1
ROBOT LANGUAGE, MANUFACTURER LISTING,
AND ROBOT ARM LISTING

Robot Language	Manufacturer	Arm
Funky	IBM	IBM
T3	Cincinnati Milacron	T3
Anorad	Anorad Corp.	Anorad
Emily	IBM	IBM
RCL	Rensselaer Polytechnic Institute	PACS
RPL	SRI International	PUMA
Sigla	Olivetti	Sigma [1 to 4 arms]
VAL	Unimation	PUMA
AL	Stanford Artificial Intelligence Lab.	Stanford
Help	General Electric	Allegro
Maple	IBM	IBM
MCL	McDonnell Douglas Corp.	Not specified
PAL	Purdue University	Stanford
Autopass	IBM	IBM

5. Validation of the solution.
6. Other techniques for verification of the solution.

In the sections that follow, the authors will discuss the robot programming principles; programming with the Alpha teach box; the internal representation of the Alpha robot; the Alpha's eighteen control functions; computer control of programming with the Alpha; electrical connections of the Alpha; and transmission rate, data format, standard format, standard interface signals, opening the port, testing the configuration, serial interface commands, and operator control of programming of the Alpha robot. Principles of industrial robotic programming languages will also be discussed.

9.2   Robot Programming[a]

The programming of a robot is, simply put, the designing of a work pattern so that the robot may perform work without human intervention. In some cases, the program also needs to cover actions of the robot that must be coordinated with the processes of interdependent machines.

---

[a] The material used in Section 9.2 is used courtesy of L. F. Rothschild, Unterberg, Towbin, New York City.

Nonservo limited sequence robots lack the memory capacity needed to store programs and, therefore, are programmed by (1) breaking down the task into a series of target positions (end points) and (2) physically adjusting the motor and setting the fixed stops to limit each movement in a given sequence. When the power is turned on, energy flows to each joint, in a specified order, moving each element so the desired target position for the end-effector results; this is repeated, as required, to produce a task cycle.

To program a servo robot, the operator breaks down the assigned task into a series of steps so that the manipulator/tool can be directed through these steps to complete the task (a program). This program is played back (and it may be repeated several times) until the task cycle is carried out. The robot is then ready to repeat the cycle. The robot's actions may be coordinated with ancillary devices through special sensors and/or limit switches. These switches, in conjunction with the controller, send "start work" signals to and receive "completion" signals from other robots or interfacing machines with which that robot is interacting.

A servo robot can be "taught" to follow a program, which, once stored in memory, can be replayed, causing the controller to be instructed to send power to each joint's motor, which, in turn, initiates motion. The feedback system and controls of a playback robot allow it to gage and direct the motions for the task (i.e., along a path) for relatively *smooth movements at variable speeds.*

To program a servo robot to perform a given task, the operator must be skilled in performing that task. There are several methods currently in use. The method employed depends on the manufacturer's specifications, the level of software basic to the control system, and the robot's computing/memory capabilities (which determine how data can be recorded). Teaching typically involves on of the following methods: walk through, lead through, or plug-in.

With the walk-through method, the operator releases all brakes and manually moves the manipulator through all positions required to do the task. The controller "remembers" the coordinates of all the joints for each position. Teaching may be done at a speed different from that speed needed for real-time operation (i.e., playback may be set at other speeds, allowing for different cycle times). This method requires minimal debugging, allows for continuous-path programming, and requires minimal knowledge of robotics. However, a thorough understanding of the assigned task is a prerequisite, and editing requires reprogramming from the error point.

Teaching with the lead-through method does not require that the operator physically move the manipulator; rather the manipulator is remotely controlled by either a computer terminal or, more commonly, a teach pendant—a device similar to a remote control box with the additional capability to record and play back stored commands. The teach pendant is plugged into the controlling computer during programming (the on-line method), and the operator then presses the appropriate buttons to position the arm, with motion being in small increments for precise positioning. When the correct position is achieved, the operator pushes a switch to tell the computer to read and store positions for all joints. This is repeated for all taught positions.

The lead-through method is often employed to program discrete points in space (through which the end-effector is required to pass) and is most commonly used for point-to-point robots. The teach pendant is most commonly used for heavy-duty robots and in those lightweight robots that have sophisticated control systems.

There are more advanced systems that allow for the movements and end points to be recorded in an *unspecified* order. This enables new programs to be created by calling out the points in a sequence that differs from the original order of input. This facilitates programming and editing. These systems also allow for defined velocity and acceleration or deceleration between points. However, these advanced systems have an inherent danger; the path resulting from a new sequence of movements may bring the end-effector in contact with nearby machinery—to the surprise of the programmer or operator.

The plug-in method is also an on-line system. The robot operates via a pre-recorded program (i.e., without manual intervention), with the program sequence having been set by manually inserted plugs in a plugboard or a stepping drum. Then, using punched cards or tape, the appropriate electrical connections are made, and those connections allow the plugboard and related sequence of motions (which have been programmed onto the board) to be directed to the mechanical workings of the manipulator. Hence, a circuit, necessary for controlling the motion of the manipulator, is created. This plugboard, with either the card or tape method, allows for faster reprogramming and, relatedly, synchronization and decision making based on external sensors. Applications of the plug-in method using decision making include orientating (i.e., aligning workpieces in designated positions) for assembly operations and material-handling work using conveyors.

The most sophisticated robot control systems have a programming capability that allows for elemental decision making, a capability needed to coordinate a robot's actions with ancillary devices and processes (i.e., to interface with its environment). Branching is the ability of the software to transfer control during program execution to an instruction other than the next sequential command. At a specific point in a task cycle, the robot will be programmed to anticipate a branching signal—a special electrical signal sent to the controller by a designated internal or external sensor. If such a signal is received, the program will follow a predetermined path or function (*branching*). If no signal is received, the program will continue to follow the main path. Thus, a robot interacting with a group of machine tools will perform a given sequence of operations, depending on which steps have been completed. For example, after a raw part is loaded onto a press (step 1), the program will look for a branching signal. If the signal is received, the program will branch to a pause, causing the robot to wait (step 2) while an ancillary machine works on that part. After the machine has completed the prescribed work, an external completion signal is sent to the controller by a sensor located on that ancillary machine. Then the robot is directed to take the part out of the press (step 3) and transfer it to another machine (step 4). Decision making can also be used to correct an operational bug. For example, a program may have a branch to a taught subprogram for releasing a jammed tool.

9.3 How to Program with the Alpha Teach Box[b]

The hand-held teach control (see Figure 9.1) performs most of the same functions as do the teach controls of large-scale industrial robots. To provide a wide range of command options yet keep product cost to a minimum, keyboard *overlays* that allow the same set of keys to provide four different kinds of functions are incorporated. Rather than use an expensive alphanumeric display to indicate which overlay is in use, a simple system of color-coded

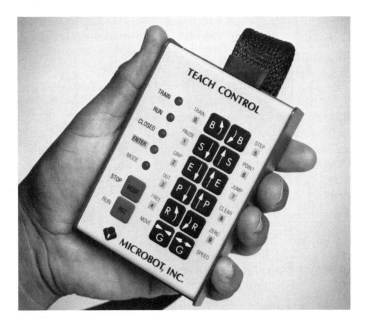

FIGURE 9.1. Teach control for the Alpha Microbot Robot System. Courtesy of Microbot, Inc.

[b] Sections 9.3 to 9.5 are from *Microbot Alpha, Vol. II, Programming Guide* (preliminary), by John W. Hill, Ph.D., Vice-President, © Oct. 1982 and are used with permission and courtesy of Microbot Inc., Menlo Park, Calif.

lights and key labels was developed. As an example, when the red MODE key is pressed, the robot is set to control mode, the red MODE light goes on, and the words printed in red (TRAIN, STEP, PAUSE, RUN, etc.) apply to the keys.

Pressing certain keys (PAUSE, OUT, POINT, HOME, JUMP, or SPEED) will cause the robot to be set to enter mode, and the yellow ENTER light to go on. When this light is on (and the red MODE light is off), the yellow numerals next to the keys apply, and numerical values can be entered. When entering values, pressing the REC button will clear the entered value, allowing the correct value to be entered. Pressing the MODE button terminates ENTER mode. When entering a value of zero, MODE can be pressed without pressing the 0 button. Pressing the AUX button when the MODE light is on also causes the yellow ENTER light to go on, but without turning off the MODE light. This is the auxiliary mode, and it is used for the FREE and ZERO functions, to control the auxiliary motors, and to record nonaccelerated moves.

Finally, the labels printed on the keys themselves (B, S, E, P, R, G, and REC) apply when the teach control is in TRAIN or MOVE mode.

9.3.1   Internal Program Representation

Prior to a discussion of the use of the teach control, an understanding of the internal program representation must be gained. All program steps (0 to 226) are 18 bytes long. The first byte is the opcode. If the opcode is null, then a null program step is present. When playing back a recorded program, the null program steps are executed. However, since the step is null, the robot does not move. In the case of a program step that moves the robot, the next byte contains the speed. The remaining 16 bytes contain the amount that each motor is to be stepped. These are double-precision integers for the base, shoulder, elbow, left wrist, right wrist, gripper, X auxiliary, and Y auxiliary motor drives, respectively. In the case of jumps, pauses, and so on, only some of the bytes are used. The values in the remaining bytes are ignored.

9.3.2   The Eighteen Control Functions

There are 18 different control commands that can be given by means of the teach control.

1. MODE and STOP.   Pressing the MODE button always stops the robot. After pressing MODE (STOP), the robot is set to control mode, the MODE light will be on, and all other lights are off except the CLOSED light, which reflects the status of the grip switch. The same is true if the unit is being operated from a host computer through the RS–232 serial port or between values when entering a two-number command.

2. TRAIN. TRAIN mode can be selected by pressing the TRAIN key when in control mode (MODE light on). Once the teach control is in the TRAIN mode, the joint-control (B, S, E, P, R, and G) keys and REC key can be used to manipulate the robot and record positions. Recording a position records the position of all six arm motors, as well as those of the auxiliary motors.

Up to 227 steps can be programmed; these steps are internally numbered 0 to 226.

Note: Pressing the REC key when in TRAIN or AUX modes overwrites the current program step and *then* increments an internal sequence pointer so that the Alpha is ready to record the *next* step.

3. RUN. To run a program, first press the MODE key to exit from TRAIN; then press RUN. To stop a program while it's running, press STOP. The MODE light will go on, and RUN can be pressed again, or TRAIN, or any other control key. Pressing the RUN button while the Alpha is running will also stop the program, but only after the program step being executed is completed.

It should be noted that, in order to simplify using the teach control, a written record should be kept of each program step. This is particularly important when programming jumps or editing programs.

Table 9.2 shows such a written record of a pick-and-place program. In this program, the robot moves over an object, lowers, grasps the object, moves it to another location, and then sets it down and releases it.

It is usually best to first record all moves at low speed,* changing them later to the desired playback speed. To change a recorded position or speed, see the teach control STEP command. To copy a recorded position, see the teach control POINT command.

TABLE 9.2
LISTING OF PICK-AND-PLACE PROGRAM

Step	Position
0	Home position [gripper open]
1	Move right and down
2	Close gripper
3	Move up and left
4	Move down
5	Open gripper
6	Move up [to clear object]
7–226	[Null]

Source: Microbot, Inc.

---

*It should be noted that a speed of zero breadth facilitates precise positioning.

4. PAUSE. If a pause is desired in a program, then:

   a. Press the MODE key (if necessary).
   b. Press the PAUSE key. The yellow ENTER light will come on.
   c. Enter a number (0 to 225) corresponding to the number of seconds the arm is to pause. If an error is made in the numerical entry, it can be erased by pressing the REC key while the ENTER light is still on. Then enter the correct value.
   d. Press the MODE key again. This terminates the ENTER mode.

The PAUSE command is saved as a program step, and the sequence pointer is incremented.

5. SPEED. This command lets the speed of the arm be changed by causing all subsequent steps and manual teach control motions to be executed at the commanded speed. The SPEED command does *not* get recorded as a program step. To change the speed:

   a. Press the MODE key (if necessary).
   b. Press the SPEED key. The yellow ENTER light will come on.
   c. Enter a number from 0 to 15. Zero is the slowest speed and 15 is the fastest. If an error is made in entering the SPEED value, it can be erased by pressing the REC key while the ENTER light is still on. Then enter the correct value.
   d. Press the MODE key again.

The correspondence between the speed numbers and the number of steps per second of the drive motors is given in Table 9.3.

TABLE 9.3
STEPPING RATES FOR THE SPEED COMMAND

Speed No.	Steps per Second	Speed No.	Steps per Second
0	28	8	600
1	50	9	720
2	86	10	900
3	129	11	1029
4	200	12	1200
5	300	13	1440
6	400	14	1800
7	514	15	2400

Source: Microbot, Inc.

There is a maximm speed at which a motor can be driven before it will start to slip. For the worst-case configuration (arm fully extended, therefore requiring maximum torque), the highest speed without slipping depends on the load the arm is carrying.

Under certain conditions, robots can be operated at higher speeds without slipping. In particular:

a. If shoulder, elbow, and wrist all descend, the robot may be lowered at a higher speed even if it is carrying a load.

b. The hand may always be closed at high speed until the hand closure contact point is reached. However, once the grip has closed, it is best to operate at a lower speed in order to build up gripping force without motor slippage.

6. STEP. Once all or part of a program has been recorded, it is often useful to move the robot through the program one step at a time. The STEP command accomplishes this. As with the other control functions, first the MODE key is pressed (if necessary), followed by STEP. This will execute the next programmed step.

Note: Null steps are *not* ignored when using the STEP and RUN buttons. So to move the robot to step 0 from step 1 of a program with no jumps, the STEP button must be pressed 226 times.

The STEP command can be used for program editing. To change an already recorded robot position, simply STEP through the program until the position to be changed is reached. Switch over to TRAIN mode, move the robot to the correct position, and then press the REC key. This overwrites the old position.

This editing procedure works because the STEP command increments the sequence pointer *before* executing or completing a program step. In other words, when the STEP command brings the robot to a given position, the sequence pointer remains pointing at the program step for that position even while the robot position is being changed. Changing the position simply changes the contents of that step location in memory.

Note: If an already recorded step is changed using the above procedure and then STEP is pressed, the robot will move to the next recorded position as usual, but this time *without* incrementing the sequence pointer. This is because the REC command already did the incrementing. A similar situation occurs when stepping through a JUMP command or when the STEP or RUN commands are used immediately after stopping or a POINT command.

The POINT command provides an alternative method of accessing program steps for editing.

The STEP command can also be used to change the speed of a previously recorded step. The procedure is as follows:

a. STEP to the sequence step to be changed.
b. Set the SPEED to the new speed.
c. Record the new speed by using TRAIN and REC (or other appropriate commands).

7. JUMP. This command allows sophisticated robot programs to be written, for it allows for *conditional branching.* When the JUMP key is pressed, the yellow ENTER light comes on, and *two* values are entered (separated with the MODE button).

• The first value represents the *jump condition.*

• The second value is the sequence pointer step number to jump to if the jump condition has been met.

The jump conditions are given in Table 9.4.

As an example, assume the robot is to move to step 5 in a program on the condition that the grip switch is open. The procedure is as follows:

a. Press MODE (if necessary), then JUMP. The yellow ENTER light will go on.
b. Press MODE (pressing 0 in this case isn't necessary since the entered value is zero).
c. Press 5.
d. Press MODE.

Or if an unconditional JUMP to step 23 is desired, then JUMP 24, 23 is used.

Note: When STEPping through a jump command, the usual incrementing of the sequence pointer is slightly modified. For example, assume stepping to a sequence step recorded as JUMP 24,7. This unconditional jump will set the value of the sequence pointer to 7. If STEP is pressed again, step number 7 will get executed, but the sequence pointer will *not* be incremented. This is because, if the pointer were to be incremented *first,* as it usually is with the STEP command, then step number 7 would be skipped, and step number 8 executed instead. On subsequent pressings of the STEP key, the sequence pointer will be incremented first as usual.

8. POINT. The POINT command is similar to an unconditional JUMP. For example, POINT 12 means set the sequence pointer to step 12 of the program and proceed from there. However, unlike the JUMP command, *POINT does not create a program step.* POINT is used simply to immediately set the sequence pointer to a given program step. It can be invoked even in the middle of program execution by simply pressing the MODE (STOP) key first.

TABLE 9.4
JUMP CONDITION ASSIGNMENT

Condition Value	Condition Definition
0	Grip switch is open
1	User opto-input bit 1 is on
2	User opto-input bit 2 is on
3	User opto-input bit 3 is on
4	User opto-input bit 4 is on
5	User opto-input bit 5 is on
6	User opto-input bit 6 is on
7	User opto-input bit 7 is on
8	User opto-input bit 8 is on or input bit 8 [uncommitted] is low
9	Input bit 9 [uncommitted] is low
10	Input bit 10 [uncommitted] is low
11	Operator control RESET button not pressed
12	Operator control program select sw. 1—not selected
13	Operator control program select sw. 2—not selected
14	Operator control program select sw. 3—not selected
15	Operator control program select sw. 4—not selected
16	Base home sensor—home is low
17	Shoulder home sensor—home is low
18	Elbow home sensor—home is low
19	Upper home sm home sensor—home is low
20	Lower sm home sensor—home is low
21	X–Aux home sensor—home is low
22	Y–Aux home sensor—home is low
23	Pad E6 is low
24	JUMP ALWAYS

Source: Microbot, Inc.

Pressing the POINT button when in control mode sets the robot to ENTER mode and the yellow ENTER light turns on. The sequence step pointer is then set to the entered value.

One of the most useful applications of the POINT command is program editing. Instead of using the STEP command to get to a program step that is to be changed, merely POINT to the corresponding program step number. Then press TRAIN (or other appropriate command) and record the new program step.

Reminder: The STEP command increments the pointer *before* executing each step. Thus, *when using STEP for editing, the program step being changed is the one the Alpha just executed.* However, *when POINT is used for editing, the program step changed is the one about to be executed.*

The STEP command can be used to verify that the proper step is being pointed to after using the point command, because the step command leaves the sequence step pointer pointing to the step just executed except in the case of a JUMP step.

The POINT command can also be used to execute multiple programs stored in memory. One could, for example, have three distinct programs recorded as shown in Figure 9.2. The programs can be isolated by means of unconditional JUMP commands.

> Step 15 would be: JUMP 24,0
>
> Step 31 would be: JUMP 24,21
>
> Step 38 would be: JUMP 24,32

To execute program 2, all that would need to be done is to

> POINT 21

prior to pressing the RUN key. Similarly,

> POINT 32
>
> RUN

will execute program 3.

The POINT command has still another application. In creating a program, sometimes it is desirable to make an exact "copy" of a program step, for example, to program a step that causes the robot to move to a position it already achieved elsewhere in the program. To do this:

a. STEP through or POINT and STEP the program until the desired position is achieved.
b. POINT to the program step to which that position is to be copied.
c. Press TRAIN, then REC.

This duplicates the desired position at the desired program step.

9. CLEAR.   This command clears all recorded robot positions and operations from program memory, and sets the sequence pointer to step 0.

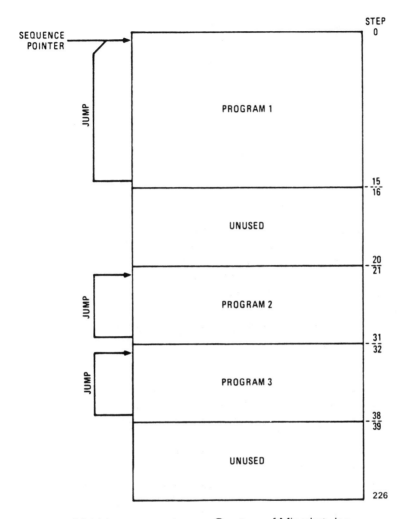

FIGURE 9.2. Multiple program storage. Courtesy of Microbot, Inc.

To operate the CLEAR command:

a. Press MODE and hold the key down.

b. Then press CLEAR at the same time.

After the MODE light comes back on, the EEPROM has been erased.

10. GRIP. This command will cause the gripper to close 64 steps of the motor past the point at which the grip switch is activated. This command is

different from using the G keys in TRAIN mode. The G keys will simply command the fingers to open or close to a particular spacing, regardless of whether the hand is holding an object. The GRIP command, on the other hand, will close on an object and then build up 2 lb of gripping force regardless of the size of the object. Thus, the GRIP command will be useful when picking up a variety of objects, whereas the G keys might be the better choice when it is desired to sense whether a particular object is present.

11. MOVE. This command activates the joint control keys as in TRAIN mode, but does not change the internal absolute position registers or allow a position to be recorded. After pressing the MOVE button while the robot is in control mode, all lights are turned off and remain off until move mode is terminated with the MODE button.

The MOVE command proves useful in moving the arm to a known position in the event of motor slippage or mechanical interference from an external obstacle. Note that this does not change the recorded program in any way; it simply lets the program continue with the arm in its correct position. For precise position control, the speed should be set to a low value (0 to 4) before using the MOVE command.

When you subsequently RUN the program, all subsequent settings of the internal position registers will now be associated with the recalibrated arm position.

12. OUT. This is the command that lets the user turn output signals on and off based on arm positions achieved or conditions met. It can also be used to turn on and off various lights on the hand-held teach control and the operator control.

When OUT is pressed, the yellow ENTER light will come on, and two numerical entries must be given (pressing the MODE key in between). The first entry is an output number (Table 9.5), and the second is 0 or 1. (In the case of the teach and operator control lights, 0 indicates off and 1 indicates on.) The output numbers are as shown in Table 9.5.

Note: The pneumatic gripper (Opto output 0, bit 15) cannot normally be used as a general-purpose output because of its action with the gripper position.

A sample program is listed next:

Step No.	Operation
0	OUT 14,1
1	PAUSE 1
2	OUT 14,0
3	PAUSE 1
4–226	[NULL]

TABLE 9.5
OUTPUT NUMBER ASSIGNMENT

Output Value	Output Controlled
0	Pneumatic gripper [Opto output 0, 15]
1	Opto output 1 [14]
2	Opto output 2 [13]
3	Opto output 3 [12]
4	Opto output 4 [11]
5	Opto output 5 [10]
6	Opto output 6 [9]
7	Opto output 7 [8]
8	Uncommitted output 1
9	Uncommitted output 2
10	Uncommitted output 3
11	Uncommitted output 4
12	Teach control TRAIN light
13	Teach and operator control RUN lights
14	Teach control ENTER light and operator control reset light
15	Teach control MODE [light] and operator control stop light

Source: Microbot, Inc.

When this program is RUN, the yellow ENTER light will blink on and off at 1-s intervals.

13. HOME. The home command is used to initialize the robot to a known position and to compensate for motor slippage. After pressing the HOME button, the robot is set to ENTER mode and the yellow ENTER light turns on.

If the MODE button is pressed immediately or a 0 is entered, the robot homes without recording a program step. If a value from 1 to 255 is entered, the robot homes and a program step is recorded.

When executing a recorded home step, the robot checks the recorded entered value with a counter that is the number of times since the last home was executed. If the recorded entered value is less than or equal to the counter, the home function is performed and the counter is reset. The current teach control speed is recorded and is used as the speed for homing.

For example, when in control mode, pressing

a. HOME
b. MODE

will cause the robot to home without recording a sequence step. Pressing

   a. HOME
   b. 1
   c. 2
   d. MODE

will home the robot and record a step that, when run, will home every 12 times through the program.

    More than one home step may be recorded in a program. Each time a home step is executed, the counter is incremented. This can be used, for example, in a program with two parts. One part could have a home step that would home every 16 times, and another part, which would be jumped to only on certain conditions, would have a home step that is executed every time.

    If the Alpha is not able to home after stepping approximately 4000 steps from the zero position, the command terminates and the robot is set to control mode. This will happen when the HOME command is initiated from the teach control, or during execution of a program, or from the serial port.

    After the robot reaches home, the internal position registers are set to zero and the home count is set to one.

14. AUX. Pressing the AUX button sets the robot to AUX mode. When in AUX mode, both the red MODE light and the yellow ENTER light are on and the red on a yellow background auxiliary overlay labels apply. AUX mode is used to:

   a. Move the X and Y auxiliary motors with the B and S button pairs.
   b. Activate the FREE function.
   c. Activate the ZERO function.
   d. Record nonaccelerated move steps.

    As in TRAIN and MOVE modes, and X and Y motors move at the current teach control speed. Pressing the MODE button exits AUX mode and returns to control mode.

15. FREE. Pushing the FREE button while in AUX mode turns off all motor currents and allows the robot to be positioned manually. After freeing the motors, the robot is set to control mode.

    In some positions, the weight of the arm can cause the robot to fall. Care must be taken when using the FREE command to ensure that the robot or external equipment will not be damaged.

    The motors will stay off until either a program is STEPped or RUN, or the corresponding button is pressed in TRAIN, MOVE, or AUX modes. Because of this, a single joint can be freed by using the FREE command and then, in TRAIN and AUX modes, locking all the joints except the one(s) to be freed.

    It should be noted that the FREE command does not change the internal position registers and therefore cannot be used to position the robot relative to absolute home.

16. ZERO. The Alpha maintains its position by a set of eight internal position registers. These registers are automatically initialized to zero when power is turned on and after a home command is completed. The ZERO command allows setting the position registers to zero at other times as well.

In addition to setting position registers to zero, the ZERO command also resets the sequence pointer to step zero and returns the Alpha to control mode.

17. Recording Accelerated Moves. Pressing the REC button while in TRAIN mode records a controlled acceleration move step. All eight absolute positions, the speed, and the opcode are saved, and then the sequence pointer is incremented. If all eight joints are to be moved during the same step, the X and Y motors must be positioned using AUX mode before recording the step in TRAIN mode.

Controlled acceleration moves should be used for most moves to get smoother motion at higher loads and speeds without motors slipping. For short moves and moves that do not change the general direction of motion, nonaccelerated moves should be considered for better performance.

18. Recording Nonaccelerated Moves. If the REC button is pressed while in AUX mode, a nonaccelerated move step is recorded. The eight absolute positions, the speed, and the opcode are saved, and then the sequence pointer is incremented. As in recording accelerated moves, the six robot joints and the X and Y motors can be repositioned before recording the nonaccelerated move step.

Smooth straight-line motions, short higher-speed moves, and continuous-path operation can be achieved with nonaccelerated move steps. For heavy loads and longer, faster motions, controlled acceleration move steps should be used to prevent motor slip.

9.4 Computer Control of Programming with the Alpha Robot

Connecting the Alpha robot to a host computer or a termianl greatly extends the unit's capabilities without losing the ability to program from the hand-held teach control.

Configuring the serial ports is the process of setting up the serial interfaces so that the computer and the Alpha can communicate. This requires the following:

1. Proper electrical connections
2. Transmission rate
3. Data format
4. Settings for standard interface signals

5. Opening the port
6. Testing the configuration

Depending on the computer being used, configuring the serial ports may not be necessary, or it may require that certain steps be taken.

9.4.1 Electrical Connections

The electrical connections on the bottom of the control cabinent show two multipin connectors, P6 and P7 (Figure 9.3). These are the two serial ports:

1. Signals that enter P7 always pass through to P6 unchanged.
2. Signals that enter P6 pass through to P7 unchanged, unless the signals are a series of characters beginning with the @ sign and terminated with a <CR> (carriage return). These signals are not passed through, but are interpreted as robot commands. One of the serial interface commands can be used to specify a different recognition character in place of the @ sign.

To operate the Alpha from a host computer, the computer must be connected to the Alpha's P6 serial port (Figure 9.4a and b). The P7 port would be used in either of the following two situations:

1. *Alpha in series with computer and other peripheral.* Some host computers only have one serial port and cannot be directly connected both to the Alpha and a peripheral simultaneously. To overcome this limita-

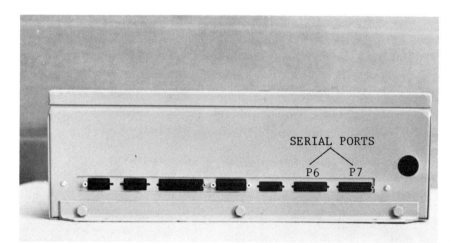

FIGURE 9.3. Two serial ports. Courtesy of Microbot, Inc.

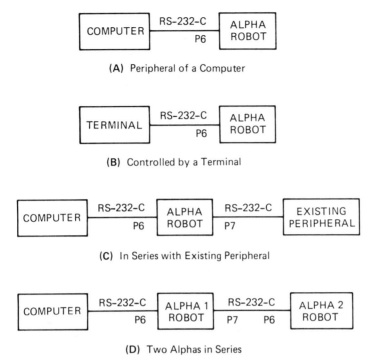

(A) Peripheral of a Computer

(B) Controlled by a Terminal

(C) In Series with Existing Peripheral

(D) Two Alphas in Series

FIGURE 9.4. Connecting the Alpha to a computer and/or peripheral. Courtesy of Microbot, Inc.

tion, the Alpha can be connected between the computer and peripheral as shown in Figure 9.4c. Signals coming from the peripheral pass through to the computer unchanged, and signals coming from the computer either pass through to the peripheral unchanged, or, if preceded by an @ sign (or other recognition character), operate the Alpha.

2. *Two or more Alphas in series.* Two or more Alpha robots can be connected in series and operated from the same computer (Figure 9.4d). In such cases, it is necessary to program each Alpha robots to respond to a different recognition character; this can be accomplished by means of the @ARM command, as will be explained later.

When using a serial line to connect the Alpha to a computer or peripheral, it is important that the transmit and receive lines be interfaced properly. On the P6 port of the Alpha, the receive line is on pin 3, and the transmit line is on pin 2. Most computers are configured with pin 3 as transmit and pin 2 as receive, so in most cases a standard serial cable will provide the proper straight-through wiring (i.e., pin 2 to pin 2 and pin 3 to pin 3).

However, it is possible that the pin assignments are reversed. This can happen if the computer has been configured to behave as a terminal, in other words, if it has been interfaced to another computer. This is because, in order to connect two computers together, the transmit and receive lines must be crossed. If the host computer has been configured to behave as a terminal, the transmit and receive lines must be reversed. This can be accomplished by modifying the cable or by reconfiguring the computer internally, as per instructions in the computer's operating manual. (*Note:* On the Alpha's P7 serial port, the transmit line is on pin 3 and the receive line on pin 2. This is so the Alpha can act as a host to whatever peripheral is connected to its P7 port.)

9.4.2   Transmission Rate

The Alpha is shipped with both serial ports configured to operate at a transmission rate of 9600 baud (9600 bits per second) for both transmit and receive. You can change this rate to any of seven other commonly used standard rates by means of three switches located on the Alpha computer card (Table 9.6). These switches should be changed when power is off, since the switch settings are read by Alpha firmware on power-up only.

TABLE 9.6
BAUD RATE SELECTION

Baud	SW1	SW2	SW3
110	ON	ON	ON
150	OFF	ON	ON
300	ON	OFF	ON
600	OFF	OFF	ON
1200	ON	ON	OFF
2400	OFF	ON	OFF
4800	ON	OFF	OFF
9600	OFF	OFF	OFF

Source: Microbot, Inc.
Note: SW4 is not used.

As with the Alpha, most serial interfaces have some means of setting baud rate, either through switches or jumpers on a circuit card, or via commands that can be issued with the computer's operating system. Both the Alpha and the host computer must be configured to operate at the same baud rate; otherwise, communication between the two will be impossible.

9.4.3   Data Format

The Alpha uses the following data format:

- 8-bit data word length
- 1 start bit
- 1 stop bit
- No parity bit

Some serial interfaces have the above as their "default" format if nothing is done to configure the format. However, with others, the configuration must be changed. Consult the host computer manual to learn how to do this.

Since the most significant bit must be zero for the first @ character to be recognized, some serial interfaces must be configured to send a 7-bit data word and a parity bit that is always set to zero. This will have the same effect as an 8-bit data word with the eighth bit zero.

9.4.4   Standard Interface Signals

Serial interfaces sometimes require logic levels on certain pins to indicate the following status conditions:

1. Data terminal ready
2. Clear to send
3. Carrier detect
4. Request to send

The Alpha does not use these signals, but does pass them through when it is placed in series between a computer and a peripheral. However, when only a single computer is connected to the Alpha (or in some cases even if the Alpha is placed in series between a computer and a peripheral), the Alpha may need to be modified in order to provide these signals.

To find out if this is necessary, consult the user manual for the computer to determine whether any of the four preceding serial interface signals are required. (With some computers, the user has control over whether these signals are required. If this is the case, then configure the computer so that these signals are *not* required.) If any of these signals *are* required, then, if a peripheral is to be connected in series with the Alpha, check the user manual that came with that peripheral to see whether the peripheral supplies the required signals for transmitting and receiving data. If it does, then all should be well. If not, the Alpha must be modified.

9.4.5   Opening the Port

Opening the port refers to configuring the host computer so that commands and data are properly routed from your computer to the Alpha. If the host computer has only one serial port, there may be nothing special to do other than use the proper commands for input and output through that port. On some computers, it may be necessary to use a special routing command or a switch setting to transmit to the Alpha whatever would normally go to a printer. Some computers have several serial ports and allow a choice of transmission channels. In all cases, consult the computer's serial interface documentation to find out what is required.

9.4.6   Testing the Configuration

Once steps 1 to 5 are completed, testing the serial connection can be done by issuing an @CLOSE command. This command closes the hand until the grip switch is activated. The details of this and all other serial commands will be described in the next section, but for now it is important to know how the Alpha responds to serial interface commands in general.

When an @ sign or other programmed recognition character is received at the Alpha's P6 serial port, all subsequent characters are buffered (temporarily saved) in the robot controller unit a carriage return, < CR> (sometimes labeled RETURN or ENTER on the keyboard), is received. The buffered characters are interpreted as a robot command.

- If the intercept command is not syntactically correct, the Alpha will return a zero (in industry-standard ASCII format) followed by a carriage-return signal. This will be designated as |0 < CR > |.
- If the intercepted command *is* syntactically correct, the robot will execute the command, then return to 1 (ASCII) with (V) READ. This will be designated as |1 + < CR > |.

This sending back of |0 < CR > | or |1 + < CR > | is called *handshaking*. After *every* robot command, it is necessary to input the handshake character into the computer so the computer knows that the Alpha is finished and ready for the next command. In addition to simply inputting the handshake character, it is recommended testing whether it is a 0 or a 1 before issuing the next command. One way to do this is by means of the following subroutine.

Note: In this and all other examples in this chapter, we will assume that the host computer uses a high-level language with serial input/output procedures. Since there are many programming languages available, we will use an imaginary Pascal-like language for these examples. When programming these examples with a particular host computer, they must be converted to the language being used. The instructions INPORT and OUTPORT represent the statements that are used for transmitting information between the computer and a peripheral unit over serial lines. INPORT causes the computer to receive information, and OUTPORT causes the computer to transmit information. For example, in some implementations of BASIC, INPORT and OUTPORT are simply modified INPUT and PRINT statements. Typical commands might be INPUT # 3, LPRINT, OUTPUT 1, or WRITE (3,STR). When a robot program is entered into the host computer, be sure to replace INPORT and OUTPORT with the proper command syntaxes.

Here is a "handshaking" subroutine that can be used to test for syntax errors and completion of Robot commands:

```
BEGIN  HANDSHAKE
      INPORT  I
      IF  I = 0  THEN  BEGIN  HERROR
            PRINT  "HANDSHAKE  ERROR"
            STOP
            END  HERROR
      RETURN  1
END  HANDSHAKE
```

This subroutine or its equivalent should be used after each robot command.

To test the serial connection, the following routine can be used. (Be sure the gripper is open before running the program.)

```
OUTPORT  "@CLOSE"
HANDSHAKE
PRINT  "GRIPPER  CLOSED"
```

If a proper connection has been made, after executing the routine two things should happen: the gripper should close, and the message GRIPPER CLOSED should be printed on the host computer's terminal.

9.4.7   Serial Interface Commands

Twelve different commands can be issued to the Alpha over the serial lines. The twelve commands are as follows:

```
@STEP        @DELAY
@CLOSE       @QDUMP
@SET         @QWRITE
@RESET       @RUN
@READ        @GOHOME
@ARM         @OFF
```

9.5   Operator Control of Programming

After the robot has been programmed with the teach control or with a host computer, the operator control can be used to start, stop, reset, select, and control robot programs. The operator control (Figure 9.5) consists of a box with two lights, three buttons, and a selector switch. The three buttons are the following:

1. STOP. Pressing the STOP button always stops the robot and sets it to control mode. This button functions like, and is the same as, the teach control MODE/STOP button.
2. START. Pressing this button while the stop light is on (control mode) starts the robot program running. The START button is the same as the teach control REC/RUN button. As with the teach control REC/RUN button, if the START button is pressed when the Alpha is running, the robot will stop when the step being executed is completed.
3. RESET. When the reset button is pressed while the Alpha is stopped (control mode), the sequence step pointer is reset to zero. This has the same effect as doing a POINT 0 with the teach control. Pressing this button at any other time has no effect except if it is tested by a program with a teach control JUMP command or through the serial port with the @READ command.

The STOP and RUN lights (in the STOP and START buttons) are the same as the teach control MODE and RUN lights.

The program select switch allows selecting one of four programs or functions with teach control JUMP commands or through the serial port via the @READ command.

FIGURE 9.5. Operator control. Courtesy of Microbot Inc.

The operator control program select switch can be used to select different programs or functions. This switch can be read through the host serial port with the @READ command. The state of the switch is encoded in the < I1 > input word sent in response to the @READ. (See serial interface @READ command.)

The program select switch can also be used with the teach control JUMP command. See the teach control JUMP command and Table 9.4. To select one of four programs with the operator control, JUMP steps must be used to test the switch.

9.6 Industrial Robotic Programming Languages[c]

The robotic language is the means of communication between the user and robot. It is important to give the robot user complete and powerful control over the robot resources, yet still maintain simplicity so that the user does not need extensive training.

To meet the needs of the user, each industrial robotic manufacturer has taken a different direction in its choice of a programming language. It seems as though there will not be any agreement on a "standard" language for some time, as we see in Section 9.1.

Some of the more popular languages available are VAL (Unimation), AML (IBM robotics), RAIL (Automatix), and T3 (Cincinnati Milacron). Many companies also take standard languages such as BASIC or Pascal and modify them to allow for real-time robot control. An advantage of using a standard programming language is that many texts exist to teach the user the language at any level of expertise.

In any case, all languages give some sort of program sequence control as well as input/output to move the robot and interface it to its surroundings (in many cases conventional forms of I/O, i.e., mass storage, are nonexistent). One of the more robust robot languages is RAIL. RAIL provides extensive I/O, many data types, sequence control, and motion commands.

The basic function of a language, such as RAIL, is to move the robot arm. When the robot is commanded to move all its axes to a new position, it is often expected that all the joints arrive at the new position at the same time. Furthermore, they should follow the shortest path in getting there. This is called joint-interpolated motion. If the robot arm has rotational joints and you were to look at the end-effector during this movement, you would notice that it is moving in some sort of arc. It might be more desirable to move the arm in a

[c]Section 9.6 is from *Introduction to Robotic Programming* by Damon Bruccoleri, Robotics Research Laboratory, New York Institute of Technology, and is used with permission. Also, the material on RAIL is used courtesy of Automatix, Inc., Burlington, Mass.

straight line to its final destination. This is straight-line movement. With carte-sian arms (such as IBM's gantry-type robot), this type of movement is easily ob-tained; with rotational arms, complicated coordinate transformations must be performed. What is generally done is to break the path down into hundreds of preliminary points (these points falling on the straight line) and then to move the arm smoothly through these intermediate points. If you were to look at the end-effector, you would notice that there would be many smaller arcs. The intermediate points are calculated from complicated mathematical for-mulas. By allowing these small arcs to approach some limiting value, the ef-fect of straight-line motion is achieved. When implementing the language, one has to decide whether to calculate these points beforehand or dynamically as the robot is moving. This decision will affect the speed of the robot and its memory requirements.

A point is a single location within the robot's working volume. A point can be expressed as the angular position of each joint relative to some arbi-trary zero position, or as the end-effector position and orientation in some co-ordinate system. For example, a cartesian coordinate system with the origin fixed at the robot base is common. For other robot configurations, a cylindri-cal or spherical coordinate system may be more natural. It is also desirable to be able to move the origin from the base to some arbitrary point. In this man-ner, all position references can be made relative to the workpiece, so that if the position of the piece is changed, then all references to it follow. For a six-axis robot operating in cartesian coordinates, a point defined as FEEDER might look like this:

FEEDER = |1000.0, 400.0, 300.0, 0.0, 90.0, 180.0|

The first three values are the X, Y, Z positions of the tool's tip in world co-ordinates, that is, using a cartesian coordinate system fixed at the base of the robot. They are linear dimensions expressed in units of millimeters or inches. The last three values are orientation angles, expressed in degrees, represent-ing the orientation of the tool relative to the world reference frame. In other words, since the tool's coordinate system moves with the tool itself, the orien-tation angles represent the tool reference frame relative to the world reference frame. The angles are the three rotations that would align the world reference frame with the tool reference frame. Picture an intermediate reference frame, X_i, Y_i, Z_i, aligned with the X, Y, Z axes of the world reference frame. The first orientation angle, ϕ, is created by a rotation of the X_i, Y_i, Z_i frame about the Z axis of the world frame until the X_i axis is aligned with the projetion of the Z axis of the tool onto the X, Y plane of the world, forming X_i', Y_i', Z_i'. The sec-ond of these is a rotation, θ, about the Y_i' axis, aligning the Z_i' axis with the Z axis of the tool, forming Z_i''. The last value is the angle of rotation, Ψ, of X_i'', Y_i'' about Z_i'' to align X_i'', Y_i'' axes with the X, Y axes of the tool forming X_i''', Y_i''',

$Z''_i{}'$, which are aligned with the tool coordinate system. This is how points are defined in RAIL.

A path not only specifies an ending point, it also conveys information on how to get there. A path can be constructed from a connected series of points. When the robot is commanded to move along a path, it will move smoothly from the start to the finish, without stopping at any intermediate points. In some languages, the robot may not actually move through these intermediate points and may not even be guaranteed to come close to it. In practice, these intermediate points are usually used to move the robot around objects in its path, and the accuracy obtained is usually sufficient. Another name given the intermediate points is "soft points." Languages can also provide a number of preprogrammed paths. Cartesian straight-line motion, joint interpolation, continuous path, and simple accelerate, cruise, decelerate paths are quite common.

Many of the painting robots control their path by the "teach by showing" method. In this method, the computer frees the robot joints so that an operator can physically grab the end of the arm and lead it through the required motions. While the operator is moving the arm, the computer is taking samples of the arm's position. Typically, they sample about 80 times per second or more. There is usually a limit to the length of the teach session since computer memory can rapidly become filled with the samples points. When the computer is commanded to enter its playback mode, it simply commands the motors through its sequence of stored points. Another name for this is continuous-path control.

Other methods for generating points are the teach-box and off-line programming discussed in Sections 9.3 through 9.5, and more recently CAD–CAM systems. The teach box, a hand-held device, is connected to the robot's computer, with the most elementary types simply providing switches on it for moving joints of the robot, and a button possibly labeled RECORD to learn that position. The teach box may include other visual indicators and switches for point editing. In fact, some are so sophisticated that whole programs can be written, debugged, and run from them. Off-line point generation might, for instance, be the user typing into the terminal the previous definition of FEEDER. If this point is slightly off in the Z direction, the user would, at the terminal, edit the Z coordinate. CAD–CAM is just beginning to be a source of points. This is the most sophisticated of the three and can include modeling of the robot application on the CAD–CAM computer and having it calculate the best points and paths. If, for example, 5000 two-inch widgets were needed instead of the 5-in. stock piece, then the computer would calculate the new robot positions, download these to the robot, start the manufacturing process, monitor the manufacturing process, and adjust if necessary. All this is just beginning to happen though and represents the state of the art. It is clear that future robots must include the ability to interface to CAM. A typical RAIL program is shown in listing 1.

```
WHILE  CYCLESTOP = = OFF  DO
       BEGIN
              MOVE  SLEW  HOME
              WRITE ['Please clamp next part in fixture,']
              WRITE ['and press "WELD" button to start welding.']
              WAIT  UNTIL  USERINPUT = = ON
              APPROACH  25  FROM  SEAM !
              WELD  SEAM1  WITH  SPEEDSCHED  [3],  WELDSCHED [2]
       END
```

LISTING 1. Sample RAIL program to weld a part upon activation of user input switch.

The WHILE statement sets up a loop and in this case will execute the statement's between BEGIN and END until the Cycle Stop switch on the robot's front panel is pressed. The first statement in the loop moves the robot to the HOME position with joint-interpolated motion. Home is where the positions of all axes are zero. For the Automatix AID 600 robot, this position is with the Y-and Z-axes retracted and the X-axis to the left. By modifying the MOVE command with the SLEW statement, joint-interpolated instead of straight-line motion is activated. The next two statements write a message to the terminal (messages can also be written to the display on the teach box). The next statement waits until the USERINPUT button is pressed. The USERINPUT button is a button on the robot's control panel. On this particular panel the button was relabeled with the word "WELD." The next statement is a bit more complicated. Many times it is desired not to directly move the robot to a taught point, but slightly before it so we can more smoothly approach the point. It is also desirable to have the tool correctly oriented at this point. The APPROACH command does this and, in our particular case, moves the robot 25 units away from the taught position SEAM1 in the direction toward the negative Z axis of the tool. If SEAM1 were defined as

SEAM1 = [1000.0, 1500.0, −300.0, 0.0, 0.0, 0.0]

then the current position would be [1000.0, 1500.0, −325.0, 0.0, 0.0, 0.0.]. Notice that this is so only because the tool coordinate lines up with the world coordinate; if the −Z direction of the tool did not line up with that of the world coordinates, then the X and Y positions would also be altered. The next command directs the robot to weld a specified point. If a path were specified, then the path would be welded. In addition, the command turns the welding arc on and then off at the end of the cycle. There are 21 speed schedules available to the user. Each schedule contains a value for the desired speed of the robot. The user can redefine any of the schedules to any value ranging from 0.001 to 2000 mm/s. There are 20 weld schedules available to the user, and they define certain welding parameters such as wire feed rate, welding voltage, predwell time, crater fill time, gas flow times, and so on.

An important consideration is if the computer's instruction pointer is synchronous or asynchronous of the robot's movement; in other words, is the robot actually executing the instruction the computer is executing? AML allows synchronous as well as asynchronous robot operation. Asynchronous operation becomes advantageous when several moves must be performed; then some time-consuming mathematical function must be calculated. While the robot is still moving, the computer can be making the calculations. If, as in our previous example, inputs were being tested, then we must be careful that these are tested in synchronism with the robot. Picture the robot executing motion commands while the robot is testing the CYCLESTOP switch 100 instructions ahead of the robot. If the operator had pressed the switch, the robot would still execute the remaining hundred motion command. RAIL answers this by only providing synchronous operation. AML provides commands to have the computer wait for the robot to exhaust its "queued up" commands.

Let's take as another example the program in listing 2, which will display the robot's status at the terminal.

```
FUNCTION  ROBOT_STATUS
    BEGIN
        WRITE ['Current robot status:']
        DISPLAY  HERE
        DISPLAY  SPEED
    END
```

LISTING 2. RAIL subroutine to display robot status to terminal.

This demonstrates the ease with which the RAIL language can be expanded. With this subroutine loaded into the computer, all the programmer now has to do is type ROBOT_STATUS as a command, and the robot's current position (obtained from the RAIL predefined variable HERE) and default velocity (obtained from the predefined variable SPEED) are displayed at the terminal. In designing a language, this expandability is important so that future hardware can be easily interfaced to the computer software. An example of this is the ease with which vision was interfaced to the AUTOMATIX computer running RAIL. It may be possible to interface a camera to a computer, but how do we write programs that will perform actual work using that camera? RAIL was expanded to include a collection of built-in functions and variables for controlling the Autovision II System. These include functions to:

1. Take pictures (i.e., store a video frame into the computer's pixel buffer).
2. Process the picture using:
   a. Connectivity analysis: used for pattern recognition.
   b. Area counting: This is very fast because it uses special hardware in the controller computer. It is intended for time-critical applications

where inspection or recognition can be performed just by looking at the area of parts.

   c. Gray-scale processing: used for color detection, texture detection, inspection of indentation and extrusion.

3. Change and display the contents of the picture memory.

4. Set windows of interest in the picture memory.

The program in listing 3 is an example of how area counting is used for parts inspection. In this example, parts are moving along on a conveyor and it is necessary to activate a light if the part is good or extinguish it if the part is bad (possibly so an operator can remove the part from the line). For this task more sensors are necessary, particularly a switch to tell if the conveyor is operating and a switch to tell when a part is under the camera. These are connected to the computer through the input port on the back of the robot's controller. The light is connected to the output port on the back of the controller. The last three statements of the program give the names CONVEYOR to the conveyor switch, PART_READY to the parts ready switch, and GOOD_PART to the light. By experimentation the operator has found that the nominal area of a bracket is 1623; the task then becomes simple:

1. Wait for a part to become ready.

2. Take a picture.

3. See if the bracket area is within some tolerance of the nominal area.

4. If it is within the tolerance, turn the light on, or else turn the light off.

5. If the conveyor belt is still active, then repeat; else end program.

```
FUNCTION BRACKET
    BEGIN
;
;   BRACKET inspects mounting brackets
'   moving along a conveyor belt.
'
        BR_AREA = 1623
        WHILE CONVEYOR == ON DO
            BEGIN
                WAIT UNTIL PART_READY == ON PICTURE
                IF OBJ_AREA WITHIN 20 OF BR_AREA THEN
                    GOOD_PART = ON
                ELSE
                    GOOD_PART = OFF
            END
    END
INPUT PORT CONVEYOR 1
INPUT PORT PART_READY 2
OUTPUT PORT GOOD_PART 1
```

LISTING 3. A RAIL inspection program for mounting brackets moving along a conveyor.

The function, BRACKET, works as follows. The variable BR_AREA is set to our nominal bracket area. If the conveyor input switch is active, then the loop following is executed and the conveyor switch is tested again. If the switch is inactive, then the function BRACKET is exited. If the loop is executed, then the program waits for the PART_READY input port to become active. A picture is then taken, and the predefined RAIL variable OBJ_AREA contains the area of the object in the picture. A test is made to see if OBJ_AREA is within 20 units of BR_AREA. If it is, we assume the part is good and activate the light; if it isn't, we assume the part is bad and deactivate the light.

This is a sample implementation of vision in a robotics system. This system as shown could only work for a simple system. However, it illustrates that a language like RAIL is self-documenting and can support a change of logic flow.

9.7   Review Questions

1. Discuss the general problem of programming in robotics.
2. Discuss the RS–232–C serial interface used in programming.
3. Discuss the hand-held teach-box control for programming.
4. Discuss computer control of programming with an industrial robot.
5. Discuss the operator control of programming.

9.8   Bibliography

Susan Bonner and Kung G. Shin, "A Comparative Study of Robot Languages," *IEEE Computer*, 0018–9162/1200.0082500, Dec. 1982.

John W. Hill, "Introducing Minimover–5" *Robotic Age*, vol. 2, no. 2, Summer 1980.

John W. Hill and Clement M. Smith, "The Microbot Teacher-mover," *Robotic Age*, vol. 4, no. 4, July/Aug. 1982.

George Katelly, "Personal Computer Networks," *Electronic Design News*, Mar. 3, 1983.

Microbot Alpha, Volume 1, Reference Guide (preliminary), Microbot, Inc., Menlo Park, Calif., Oct. 1982.

Microbot Alpha, Volume II, Programming Guide (preliminary), Microbot, Inc., Menlo Park, Calif., Oct. 1982.

Robotics Reference and Applications Manual, Minimover–5® Part No. RRA–1, revision 2, Microbot, Inc., Menlo Park, Calif., 1982.

TeachMover User Reference Manual, Part No. RR2, edition 2, Microbot, Inc., Menlo Park, Calif., 1982.

10
Robot Work Cell Environments

10.1   Introduction

The robot cell can be defined as the robot plus the accessory equipment necessary to create a particular environment. The manufacturing system where the robot or robots work is the work cell. This could be also created from a work station. Applications of work cells are applicable to the following manufacturing uses:

1. Die casting
2. Forging
3. Machine load and unload
4. Parts transfer
5. Spray painting
6. Small parts assembly
7. Finishing
8. Plastic molding
9. Welding
10. Machining
11. Electronic assembly
12. Inspection
13. Printed circuit assembly

To provide an understanding of robot cells, we shall consider applications of the Integrated Computer-Aided Manufacturing (ICAM) program of the U.S. Air Force; Unimation, Inc., Work Cells; and Prab Robots, Inc., robotic work cells.

Presently more than 15 manufacturers use turnkey-type electronic components insertion/placement robotic stations or systems. The robot arm can adjust mechanical parts or check electrical parts. Anywhere a robot is seen, it is a part of the robot or robotic cell.

10.2 Robotic System for Aerospace Batch Manufacturing[a]

The control system, composed of computer hardware, software, and interfaces, forms the heart of a robotic work cell. The need for dividing the overall control problem into a number of different forms for an integrated control system is required. Each level in the partitioned control system accepts commands from the next higher level and responds by generating ordered sequences of simple commands to the next lower levels. The control system uses sensory feedback to close control loops when appropriate.

The components chosen for the development of the robotic work cell integrated control (Figure 10.1) are as follows:

1. An Intel 80/20 computer system for cell control.
2. A Qantex 650 cartridge tape drive for mass data storage.
3. An LSI ADM 3 dumb terminal for operator interaction.
4. A Datel APP 20 A1 printer for hard copy of system status.
5. A recognition equipment OCR–A reader for automatic part identification.

Software includes robot control programs for a robotic work station and an executive program for the integrated control computer.

Sensors in the robotic system include the following:

1. Status of technology
2. Automated part identification
3. Contact and tactile
4. Optical and photoelectric
5. Safety

[a] The material in Sections 10.2 and 10.3 is exerpted from "ICAM Robotics System for Aerospace Batch Manufacturing—Task A," courtesy of Materials Laboratory, Air Force, Wright Aeronautical Laboratories, Air Force Systems Command, Wright-Patterson Air Force Base, Ohio.

ROBOT HIERARCHICAL CONTROL

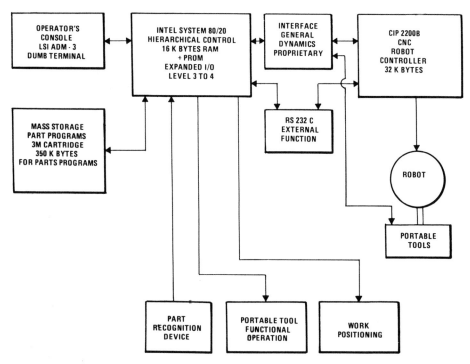

FIGURE 10.1. Robotic work cell.

10.2.1 Robotic Work Station

The work station was designed to meet the following functional requirements:

1. The station must employ a currently available industrial robot.
2. The station must be capable of drilling and routing a limited number of sheet metal parts of sizes up to 3 ft by 4 ft.
3. One worker must be capable of operating the station with minimal skill requirements.
4. Selection of operators from the shop floor must be possible.
5. Manual parts fabrication in the station must be feasible in case of system failure.
6. Hole patterns and router paths must be defined by templates.
7. Compliant end-effectors must be employed due to inherent inaccuracies of available robots.
8. Part programming must be done on line.
9. Safety sensors must be installed for protection of personnel, equipment, and parts.

10. Consistent quality must be maintained on parts, with no rejects.
11. Productivity increase must be proved to make the system economically feasible.
12. Early implementation must be made for quickest payoff.

The General Dynamics robotic work station was constructed around a Cincinnati Milacron T³ Robot. This robot was selected from those commercially available because its combination of capabilities was the most suitable for the intended applications. Considerations in making the selection were load capacity, reach or operating volume, accuracy of positioning, power and flexibility of the control system, and adaptability of the robot to use modified versions of the portable tools of the aircraft industry. The robot was installed in a laboratory environment where development could be carried out free of other constraints and pressures. A work-station concept (Figure 10.2) was established in which the robotic work station would operate with modified pneumatic tools and a part-positioning fixture to drill and rout aircraft

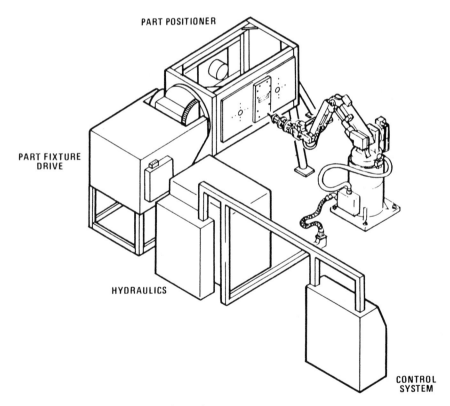

FIGURE 10.2. Robot work station concept.

panels. The technology assessment and plan of action accomplished in Phase I of this program indicated the type and extent of modification required to refine the work station for batch production capability.

With the assistance of the coalition, General Dynamics assembled, adapted, constructed, and developed system components to configure a functional, production-ready work station for drilling and routing aircraft sheet metal panels (Figure 10.3). The hardware and software components of the overall control architecture were assembled and integrated into the work station configuration, and a major effort was expended in development of software for the functions of drilling and routing. A part-positioning fixture and modified pneumatic tools were incorporated.

The robot control system was taught the coordinates in its work space and to make the movements required for each operation. In addition, the control system was taught to accept sensor signals or programmed instructions for activating the drilling and routing tools used in machining operations. Signals from safety sensors were interfaced to the control system to cause the robot to execute appropriate programmed actions to eliminate hazards. Interlock devices were also linked with the control system to ensure safety of operating personnel and others who may come within reach of the robot arm.

The experience obtained by General Dynamics in the prototype work station was beneficial in developing the power tools and fixtures for the work station. The robot, through the tool attachment mechanism, represents one interface with the portable power tools; the fixturing represents the other interface. Each tool system interacts with its power source, the fixturing, the workpiece, the robot arm, and the control system through mechanical and control interfaces. Sensors integral with the tools perform the following functions: to verify that conditions are proper before the tool may operate; to take corrective action when conditions are unfavorable; to signal when conditions are right; and to signal when the function is completed. The tools have adequate power and operating speed for clean and rapid cutting. Since robot positioning accuracy is insufficient for high-precision cutting, a smoothly operating compliance mechanism was developed for use with the power tools. The drilling and routing heads and the fixturing were developed in the General Dynamics Production Integration Laboratory (PIL) and were tested on typical F–16 sheet metal parts.

The Cincinnati Milacron T^3 industrial robot utilizes direct, electrohydraulic drives to control each of the six degrees of freedom. This robot was discussed in Section 8.15. It has a maximum reach of more than 97 in. and can position a 175-lb load at speeds of 50 in./s. The selection of this particular Cincinnati robot was based on over two years' experience with this model and its immediate predecessor, the Model 6CH.

The specific performance characteristics that make the T^3 robot acceptable for this task are as follows:

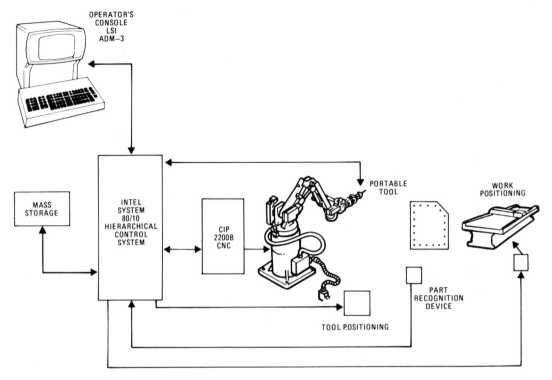

FIGURE 10.3. Production work station configuration.

1. Repeatability accuracy of ± 0.50 in.
2. Satisfactory load-carrying ability.
3. Reach parameter capable of working a 3 ft by 4 ft workpiece.
4. Sufficient memory for program storage.
5. Satisfactory control system for smooth path guidance and programming flexibility.
6. Domestic manufacture by a firm with a reputation for service.

The robotic drilling and routing work station sequence of operation is shown in Figure 10.4. Bench preparation of parts is shown by task center as well as the robotic drill/route area.

The basic configuration of the work station and the functions of the individual components are shown in Figure 10.5. A Cincinnati Milacron T³ industrial arm robot chosen for the work station has been installed in the robotics laboratory. This commercial robot was found to have the greatest accuracy for its reach and load-carrying capacity. In machining operations, the robot arm is used for positioning the drill and router systems at the workpiece. Arm positioning is controlled by the robot control computer through feedback ser-

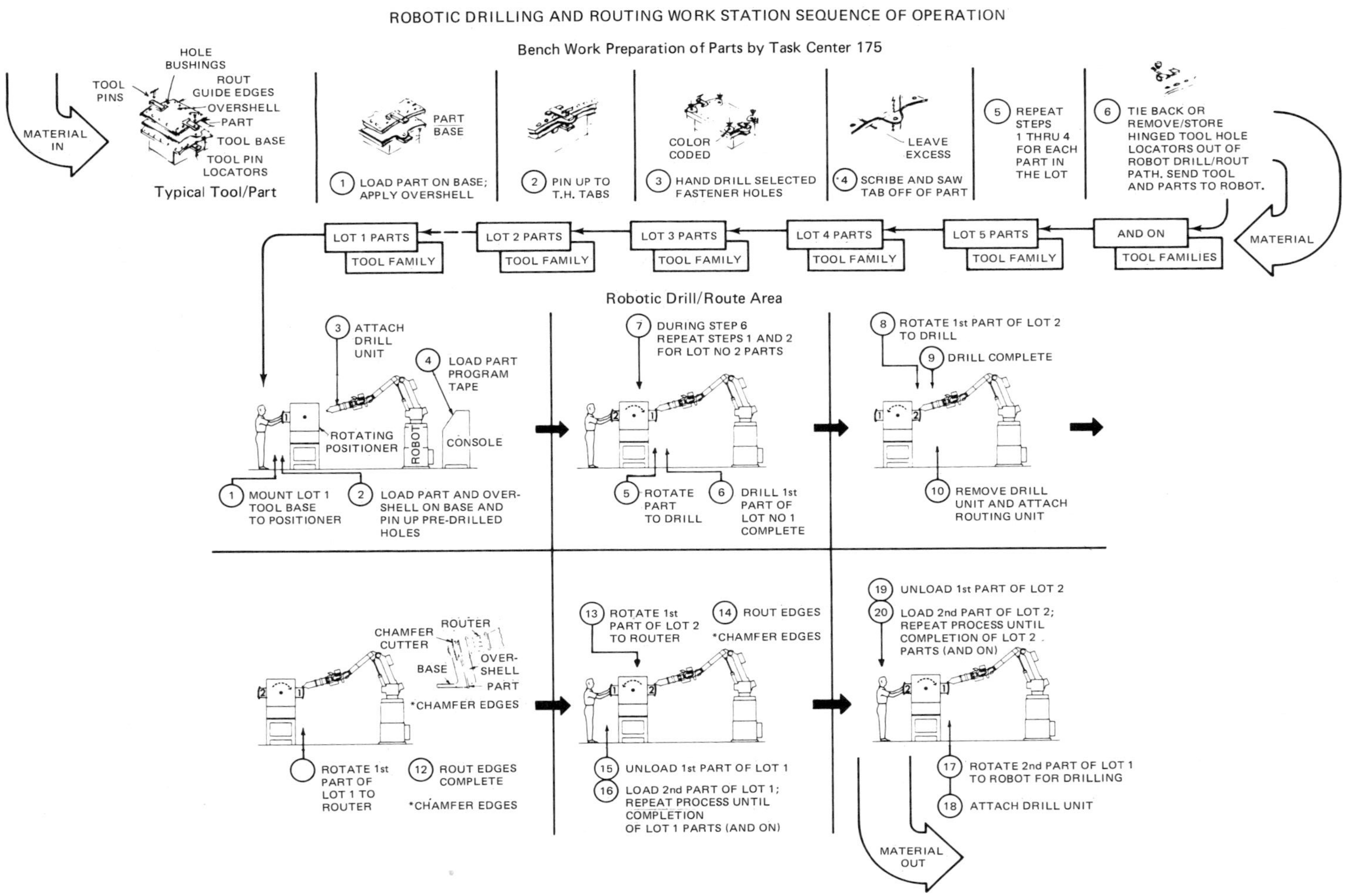

ROBOTIC DRILLING AND ROUTING WORK STATION SEQUENCE OF OPERATION

Bench Work Preparation of Parts by Task Center 175

Robotic Drill/Route Area

FIGURE 10.4. Robotic work station scenario.

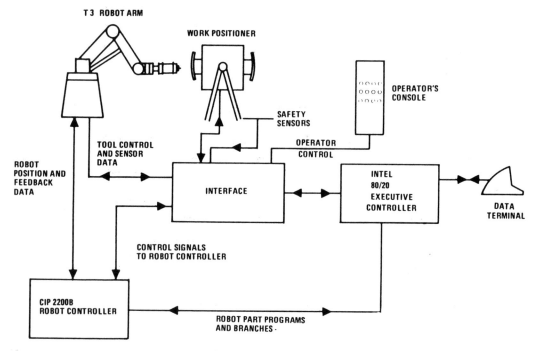

FIGURE 10.5. Robotic work station components.

vos, tachometers, and control valves interfaced to each of the six axes of the arm.

The robot control computer is a CIP 2200B computer-numerical-control (CNC) supplied with the robot. The functions of the CNC are as follows:

1. Provide control and coordination of the six axes of the robot arm.
2. Communicate with an external computer through an RS-232-C communications data link and a computer-to-computer interface.
3. Provide data point storage space for the part programs.

The CIP 2200B is equipped with a cassette loader for loading its operating system tape and part programs. This allows the robot to operate in a stand-alone mode if the integrated controller malfunctions. In the stand-alone mode, the robot controller is limited as to the number of parts it can process automatically, since data point storage capacity is only 685 points. With very careful programming, five to six small parts can be processed. Due to this limitation, the robot computer control is supplemented with an integrated control computer.

The integrated control of the station is provided by an Intel System 80/20 microcomputer. Communication between the Intel 80/20 microcomputer and the CIP 2200B is through the standard RS-232-C communications data link and a dedicated electronic computer interface. Input/output signals from the CIP 2200B minicomputer are tied directly to the Intel System 80/20 microcomputer's input/output ports through this interface. With the expanded I/O and memory in the Intel 80/20 microcomputer, the computer can communicate with other fabrication cells or with a host computer.

The Intel 80/20 microcomputer supervises the work flow in the work station, provides general record keeping, and operates the portable tools.

Two different portable tool systems, a compliant drill system and a compliant router system, are to be used in the work station. The tool systems are coupled to the robot arm through a quick-change adapter fitted to the robot arm.

The drill motor is housed in a highly compliant cage fitted with a guiding nosepiece. Used in combination with a fiberglass template with hardened guidance bushings, it forms a hole-locating system required for close-tolerance drilling of aircraft parts. This hole-locating method is capable of holding tolerances within ± 0.005 in. (0.0129 cm), even though positioning accuracy of the robot arm is no better than ± 0.050 in. (1.27 cm). The compliant drill cage also houses the sensors that detect cage extension and retraction, drill extension and retraction, and part contact.

The Intel 80/20 is interfaced to the drill system to control and monitor the drilling operation. The router is also housed in a compliant cage. Bearings are used to guide the cutter along the periphery of the template. The router motor and coolant are controlled by the Intel 80/20 microcomputer.

The part positioner for the work station is a commercially available Aronson weld positioner retrofitted with a rotating box of aluminum plate. The part positioner holds the workpiece in the working arena of the robot arm for drilling and routing, while the operator is unloading and loading another workpiece on the opposite side.

The drilling and routing templates consist of two fiberglass shells that have the contour and hole pattern of the sheet metal part. The sheet metal part is held between the fiberglass shells by clamping devices through two index holes drilled by hand. The outer fiberglass shell has hardened drill bushings for locating the drill nosepiece in the hole pattern of the part.

The operator can interact with the robot, the tools, and the part positioner through the operator console. The console consists of an LSI ADM 3 CRT terminal with a keyboard and a robot control panel for operations of the robot arm. Operations conducted from the panel include drill bit change, abort operation, tool check, interrupt, and emergency stop.

Software for execution of the drilling and routing operation and integrated station control is implemented in the work station. Maintenance programs for checkout of the hardware are also implemented.

10.3 Robotic Work Cell

10.3.1 Overview

Phase III of Task A in the Robotics for Aerospace Batch Manufacturing called for the expansion of the robotic work station into a fabrication cell. In addition to all the capabilities of the work station, the fabrication cell was designed to have the following added capabilities: automatic part identification, computer-controlled part positioning, automatic tool changing, hard-copy printout of cell status, material handling, mass data storage for parts programs, and integrated control of the total work cell.

Functional requirements for the work cell included all of the requirements for the work station, plus the following additional requirements:

1. Automatic parts positioning
2. Automatic part identification
3. Automatic tool changing
4. External storage of parts programs
5. Integrated cell control
6. Capacity to operate with little or no operator intervention
7. Parts handling equipment
8. Automatic drilling and routing of numerous parts
9. Flexibility for numerous tasks

10.3.2 Configuration

The work cell configuration consists of an automatic parts positioner, automatic part identification, and automatic tool changing.

Automatic Parts Positioner. The computer-controlled rotating parts positioner was the first component to be integrated into the system. The positioner is an Aronson Weld Positioner modified for external computer control. The positioner was designed to allow the robot to be working a part at the same time the operator is loading a new part or unloading a finished part from the opposite side. Upon completion of its current task, the robot notifies the integrated controller and the new part is rotated into the working area of the robot.

Automatic Part Identification. Part identification is accomplished by the use of a commercially available optical character reader (OCR). The OCR is interfaced to the mass data storage routines, and the part number is read into a buffer in the integrated controller. The mass data storage software then searches the magnetic tape for the proper part program. When the proper program is located, it is then loaded into the robot controller for processing.

Automatic Tool Changing.   One part of the automatic tool change feature is the tool rack. While the tool rack serves for storage of the drills and router, the automatic coupling of tools is accomplished through software in the robot controller, the integrated control computer, and sensor hardware on the robot arm and end-effectors.

10.3.3   Description of the Work Cell

The limitation of the work station (drilling and routing only a small number of parts in a batch manufacturing environment) will be eliminated by upgrading the work station into a work cell. The planned work cell will have all the capabilities of the previously described work station, plus the added capabilities of automatic part identification for handling numerous parts efficiently in a batch manufacturing environment and expanded part processing capabilities.

The basic configuration of the work cell (Figure 10.6) will have all the components of the work station, as well as the following added capabilities:

1. *Mass data storage.* A magnetic tape drive unit will be added to the station. The tape drive uses a 3M type magnetic tape cartridge that can store 2.5M bytes of data. The magnetic tape will be used to store all

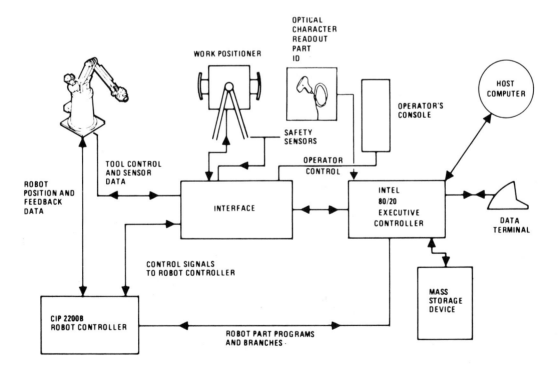

FIGURE 10.6. Robotic work cell components.

part programs. The part programs will be transferred from the CIP 220B minicomputer in the teach mode through the Intel 80/20 microcomputer to the magnetic tape.

2. *Material handling.* A material handling system for efficiently routing workpieces in and out of the cell is being considered. The material handling is most likely to be accomplished with overhead cranes and dollies, which are used throughout the factory. A balance master will be used at the outset for assisting the operator in placing the workpiece onto the positioner. The subcontracted coalition has been tasked to assist the Task A contractor in developing an efficient and economical material handling system.

3. *Software.* The integrated computer control concept for the work cell requires that software development be an ongoing task.

The robotic work cell evaluation is shown in Figure 10.7. The robotic system for aerospace batch manufacturing is seen through evaluation of the robotic cell. The end result is an integrated station control system.

10.3.4   Work Cell Operation and Evaluation

The operation of the robotic work cell is shown in Figure 10.8. The sequence is described in detail as follows:

1. The Cincinnati T^3 robot system is powered up. The hydraulic system is allowed to warm up and stabilize at operating temperature. The robot is started in the AUTO mode and proceeds to a WAIT function.

2. The Intel 80/20 computer is started and is ready to control and monitor work cell activity.

3. The operator loads the part onto the rotatable cube work positioner, as in the operation of the work station.

4. The part number is read by means of the optical character recognition device or entered at the keyboard.

5. The Intel 80/20 computer searches the tape for the part file; when found, the computer retrieves the robot program from the tape.

6. The Intel 80/20 computer initiates the rotation of the work positioner and monitors the operation for safety interrupts until completion.

7. The Intel 80/20 computer enables the robot to enter process branches and then indicates robot activity by taking it out of a WAIT state.

8. The robot requests and receives external data from the Intel 80/20 computer via the RS-232-C link by means of the EXTERNAL function.

9. The robot halts in another WAIT state.

10. The operator indicates part processing manually by depressing a switch on the control panel; this causes the robot to come out of a WAIT state.

ROBOTIC SYSTEM FOR AEROSPACE BATCH MANUFACTURING
EVOLUTION OF ROBOTIC WORK CELL

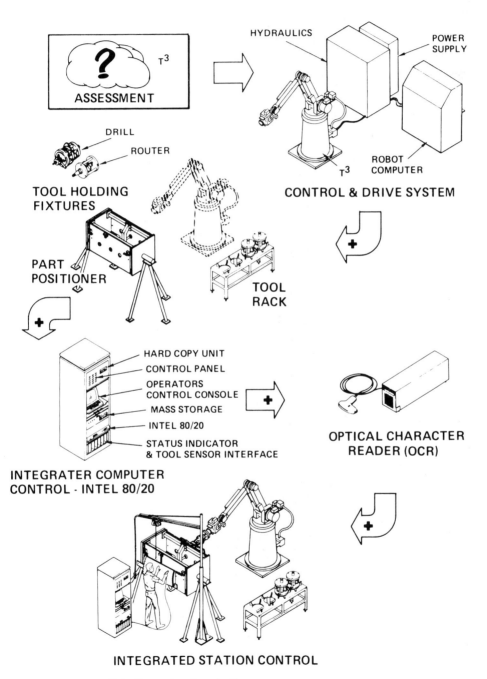

FIGURE 10.7. Robotic work cell evaluation.

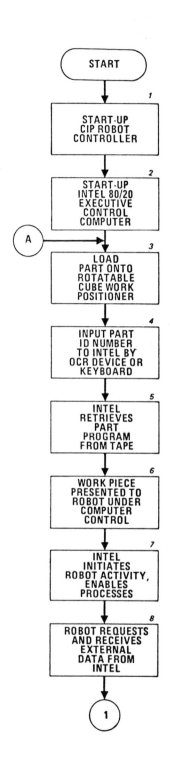

FIGURE 10.8. Work cell operation.

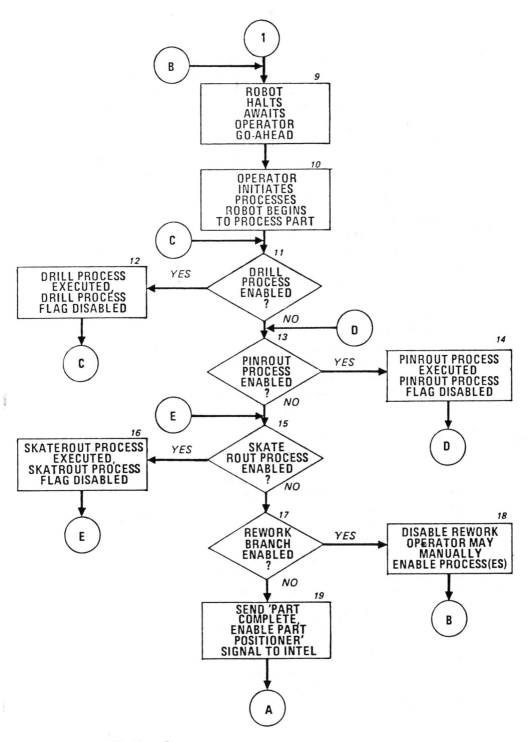

FIGURE 10.8. [Continued].

11. If the drill branch is enabled, the robot enters drill routine.

12. The robot picks up a drill from the tool rack and drills the holes under the supervision of the Intel 80/20 and with the aid of the template. Upon completion of this operation, this computer disables the drill branch. The process flow then returns to block 11 as shown and proceeds to block 13, because block 11 is now negative.

13, 14. Same as blocks 11 and 12, except a pin router is used to rout the periphery of the part. Process flow returns to block 13 and proceeds to block 15.

15, 16. Same as 11 through 14, except a skate router is used. Process flow returns to block 15 and proceeds to 17.

17, 18. If this branch is enabled, the robot enters the branch, signals the Intel 80/20 to disable this branch, and then returns to 1. The operator inspects the part. If any of the preceding processes need to be repeated, the operator manually selects one or more by depressing switches on the control panel and restarts the robot. If no process was reselected, block 19 is executed.

19. A PART COMPLETE signal is sent to the Intel 80/20. The Intel 80/20 checks the OCR buffer to see if a part number was entered (indicating a part was loaded and part number entered while the robot was working on the previously finished part). This is after the process flow has returned to entry point A (blocks 3 and 4).

10.4   Applications of Unimation Work Cells[b]

The Unimation, Inc., System Division provides manufacturers with the best utilization of their existing equipment as shown in the block diagram of Figure 10.9. In this configuration we see a robot, numerical control machiner, inspection units, thermal treatment, and inspection units fed to a work cell controller. The work cell controller consists of a terminal, log listing, and database that goes to a mainframe computer.

The Unimate robot system (Figure 10.10) used by the Xerox Corporation for a high-speed machining line is designed to produce parts for their 9200 series duplicator. The robot work cell system enables the company to reduce floor space requirements, achieve faster operating cycles, and cope with model changes and variable part demand.

Another application of Unimate robots is found in agriculture. Massey Ferguson, the agricultural equipment manufacturer, installed a flexible automation system in link eight stand-alone chuckers, shapers, and shavers. The company chose Unimate robots after learning that special loaders

[b] The material in Section 10.4 is used courtesy of Unimation,®—Inc., Danbury, Conn.

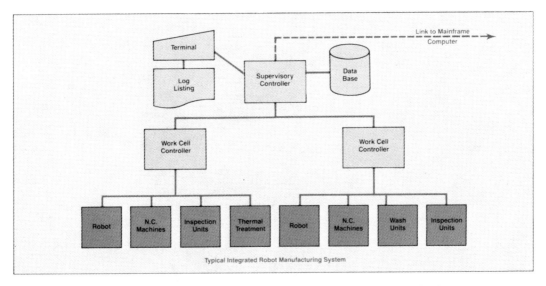

FIGURE 10.9. Typical integrated robot manufacturing system. Courtesy of Unimation®, Inc., Danbury, Conn.

(special-purpose automation) would cost more in time and money than robots. See Figure 10.11 for a robot cell system for this particular application.

The advantages of this system are at least 25% more productivity over the long term, as compared to a manual method, and flexibility (the system can be changed over rapidly to accommodate four different sizes of pinion gears).

Figure 10.12 is a typical layout of a robot cell assembly line showing the combining of robots and manual work stations. Figure 10.13 shows a 13-robot work cell system used for a total of 450 spot welds on car bodies.

A 13-robot work cell system is used by the Chrysler Corporation. Besides permitting the reduction of time and cost of model changeovers, the system

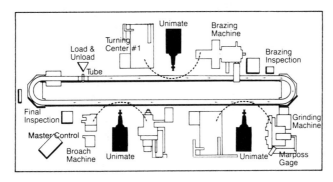

FIGURE 10.10. Unimate robot cell system for high speed machining. Courtesy of Unimation®, Inc., Danbury, Conn.

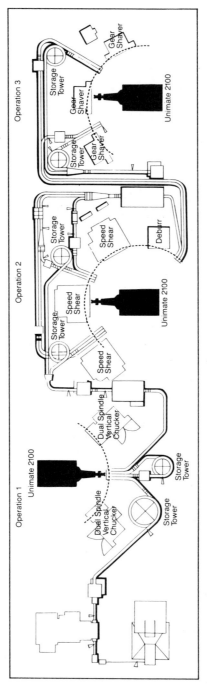

FIGURE 10.11. Unimate robot cell used by agriculture manufacturer. Courtesy of Unimation®, Inc., Danbury, Conn.

FIGURE 10.12. Typical layout of a robot cell assembly line combining robots and manual work stations. Courtesy of Unimation®, Inc., Danbury, Conn.

produces welds of greater consistency and uniformity for better joint integrity and overall product quality. The flexible system easily adjusts to variations in line rates and car body styles.

A five-robot, fully automatic investment casting robotic cell system is shown in Figure 10.14. The Precision Castparts Company of Portland, Oregon, uses work cell techniques to process large investment castings and a robotic

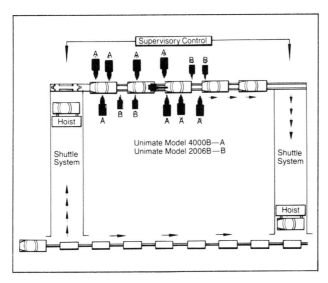

FIGURE 10.13 A 13-robot work cell system used for a total of 450 spot welds on car bodies. Courtesy of Unimation®, Inc., Danbury, Conn.

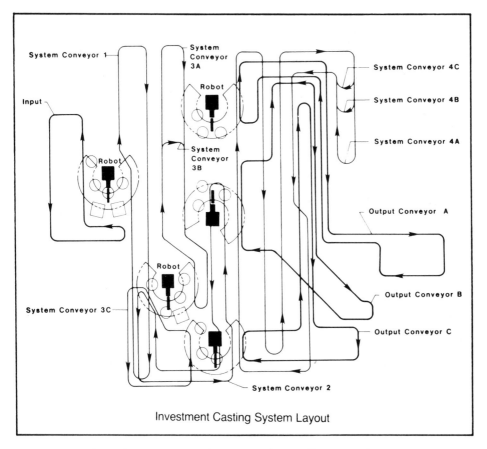

Investment Casting System Layout

FIGURE 10.14. Investment casting system layout. Courtesy of Unimation®, Inc., Danbury, Conn.

work cell. Five 4000 Series Unimate robots, conveyors, and supervising control are used in the processing technique.

The first two Unimate robots operate in a series mode, processing molds weighing approximately 150 lb through successive slurry dip and sand coating operations. The other three robots operate in parallel, receiving molds randomly from Unimate robot 2. The robots will manipulate one, two, or three molds simultaneously depending upon mold size.

Since various part and lot sizes are run through the system randomly, two Modicon industrial process control units are used to control all system activities in a master–slave arrangement. As a wax tree enters the system, an operator assigns a code number to specify the sequence of dips and sanding operations required for that particular part. This information labels the shell and assigns each Unimate robot's program to process that shell through the system. Complete production history for each part is available on print-out at the end of the line.

10.5 Application of Prab Robotic Work Cells[c]

Prab Robots, Inc., one of the leading manufacturers of robots and conveyors in the United States, presently has a new robotic work cell for full integration of factory automation and the Prab FB Robot for palletizing three automobile engines at one time. With its heavy payload carrying capability, the robot demonstrates high-speed and close tolerance repeatability for multiple parts and intelligence for performing complex palletizing routines. Three other robot work cell problems are shown in Figures 10.15 through 10.17.

To automate a combination of three mechanical forging presses, ranging in size from 2500 to 5000 tons, and to feed three 1500-ton mechanical trim presses all automatically is the problem. The solution is two PRAB robots, which are assigned to each combination of forging and trim presses (see Figure 10.15). One robot places a billet in the block die and removes the forged part from the finishing die. The robot moves horizontally to the corresponding trim press, places the part in the die, signals the press to trim, removes the part to a storage bin, and repeats the cycle.

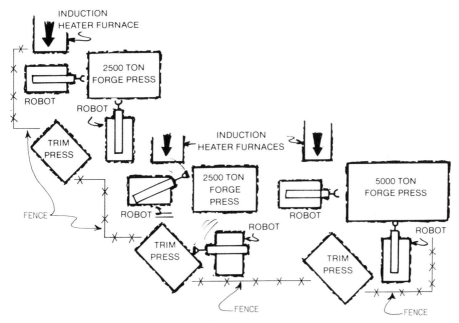

FIGURE 10.15. A robotic cell in a large forging shop in California. Courtesy of Prab Robots, Inc., Kalamazoo, Mich.

---

[c] The material in Section 10.5 is used courtesy of Prab Robots, Inc., Kalamazoo, Mich.

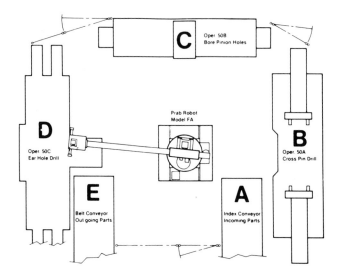

Production cell layout. Letters indicate work-flow.

FIGURE 10.16. Robotic cell layout for automating loading and unloading of truck differentials. Courtesy of Prab Robots, Inc., Kalamazoo, Mich.

While the first robot handles the secondary operations, the second robot is moving the billet through the forging operation.

With the installation of the industrial robots, this forging operation requires only one operations person, versus three utilized by the manual operation. It is a six-day, three-shift operation that produces parts ranging in size from 10 to 83 lb. Production rate is 120 pieces per hour.

Operator fatigue can hinder productivity. Manually loading 20-lb parts into drilling and boring machines required the operators to handle approximately 12 tons of castings during each shift. The physical demands of this operation led to erratic cycle times, and the machine operators had little time or energy to verify part quality.

After careful investigation, the Prab Model FA Robot was used to fully automate the laborious loading and unloading tasks. Prab's application engineers, working in close cooperation with the manufacturing engineers, determined the optimum work cell arrangement.

Three drilling and boring machines, designed for the application to handle two parts at a time, were positioned around the hydraulically-driven robot (see Figure 10.16). An indexing conveyor leading into the cell provided precise part orientation, while a simple belt conveyor with unloading fixtures was installed to receive outgoing parts.

While the parts move through the cell chronologically as indicated on the layout sketch, the robot's routine sequence is basically from D to A, because

FIGURE 10.17. The Prab Model FB Robot transfers two truck differentials through drilling and boring operations at the Eaton Corporation plant in Marshall, Michigan. This photo shows the automated work cell concept with the robot in the center and machine tools and conveyors surrounding the robot. Courtesy of Prab Robots, Inc., Kalamazoo, Mich.

the machine fixtures must be emptied before another pair of parts can be loaded.

Productivity in the work cell has increased nearly 60 %. Direct labor costs are greatly reduced. One man is assigned to supervise the entire operation during each shift, providing constant inspection and quality control. Considerable savings also result from the fact that none of the machines sit idle beyond planned cycle times.

Photocells sense proper part orientation at the incoming automatic indexing station, signaling the robot to begin the transfer sequence. Spring-actuated locator pins on the index fixtures are released with a simple, forward motion of the robot arm. A photocell on the output fixture verifies that the robot has released the finished parts, while limit switches indicate whether the hand tooling is open or closed and whether the cell's hazard enclosure safety fence has been opened.

In Figure 10.17, the Prab FB Robot transfers two truck differentials through drilling and boring operations at the Eaton Corporation plant in Marshall, Michigan. The work cell concept is shown in the center.

As robotic technology becomes more sophisticated, so will the work cell environments.

10.6   Review Questions

1. What is a robot cell?
2. Draw a robot cell and label all parts.
3. What is a work station?
4. List the component parts of a work station.
5. List the component parts of a work cell.

10.7   Bibliography

Conference records of the Robot VII 1983 Conference presented by the Society of Manufacturing Engineers. Published yearly. Records have been published since Robot I by the Society of Manufacturing Engineers.

Robots in Industry, Unimation, Inc., Shelter Rock Lane, Danbury, Conn. Newsletter published every other month throughout the year.

11

Economics and Growth of Robots

11.1   Introduction

In our society today, everything is costly. Industrial robots are also expensive. Some of the considerations to be taken up in this chapter are the following:

1. Developments needed for better use of robots.
2. The impact of industrial robotics on American productivity.
3. Competition of American robots versus Japanese robots.
4. Robots used in other areas beside factories.
5. Repair of robots.
6. Small companies being eliminated.
7. Modular approach to future robots.
8. Factory automation, including inspection, assembly, and quality control

Benefits of robotic technology include the following:

1. Ten percent to 67% improvement in productivity reported as of 1983.
2. Robot up time as high as 98%.

3. No second- or third-shift production slowdowns.
4. More effective use of expensive capital resources will result in longer life from tools and dies.
5. Lower average absenteeism in the human work force.
6. Lower cost of compliance with federal regulations.
7. Improved product quality results from fewer rejects and less rework; less scrap; up to approximately 70% improvement reported in quality.

Board-test manufacturers are now automating their products with electro-mechanical robotic board handlers that feed and retrieve boards to and from the test stand. While the majority of robots in operation today are aimed at handling heavy-duty and hazardous tasks, several companies are targeting their first product at light electronics assemblies.

Robotic applications will include the nuclear industry, agriculture, undersea activities, and construction and other areas. Rapid growth in the industrial robotic market could propel robot shipments in the United States from $315 million in 1982 to $3 billion level in 1992.

Senior and community colleges will assist in robotic development with manufacturing support now and in the future. Many senior colleges have shown an interest in robotics by generating at least one course in robotics. With the economy as it exists today, changes must be made in the price of a robot so that community colleges and others can purchase an industrial quality unit for less than $6000. Grants from the government and private sources should be of assistance to help with this problem. This probably will take several years to achieve.

11.2   Japan and Industrial Robotics[a]

The capital-intensive nature of Japan's robot facilities means that in many cases high volumes should be produced in order to justify the investment. Thus, Japan has produced a higher volume of robots and software, bringing down the unit cost to produce a given item, and then exported those items heavily.

Growing trade friction has meant that Japan must start to increase local production (and also that the United States must increase its production in Japan). Moreover, mergers, acquisitions, licensing agreements, and joint ventures between Japanese and U.S. robotic companies are likely to increase as both countries recognize the strengths and weaknesses of one another. That is, interdependence for the mutual benefit of both countries, such as we have already seen in the robot industry, should be a growing phenomenon.

---

[a]The material in Sections 11.2 to 11.5 is used courtesy of Laura Conigliaro and Prudential Bach Securities, New York City.

The types of industries upon which Japan is likely to have an impact have been changing. In the 1960s, it was textiles, followed by steel and color television sets in the early 1970s, and automobiles in the late 1970s. In the future, Japan will be emphasizing industries that are related to the microchip—what Japan calls the "new wave" industries. These include computers, office automation equipment, integrated circuits, industrial robots, optical fiber communication, and, though not growing out of the microprocessor field, the area of life sciences. These also happen to include those industries where the United States has built great strength. This should increase the necessity for U.S./Japanese and Japanese/U.S. local capital investments and mutually beneficial agreements in robotics. It also means that U.S. companies, to be successful, are probably going to have to be smarter and somewhat more innovative while not ignoring ways in which to use to their advantage what Japan has to offer.

Japan also will increasingly take advantage of the fact that its location in Southeast Asia is within the largest population area, the area where both GNP and population are growing most rapidly. Within that geographical area, it is the most highly industrialized nation. This should, over time, give Japan a large new market for growth as it targets developing countries.

With this "new wave" sector as regards manufacturing technology whereby integration is the most important characteristic and goal, Japan has already been developing a wide spectrum of manufacturing technology experience (see Figure 11.1).

Many different sources are aware of the fact that Japan is at least a few years behind the United States in software. Japanese companies by and large are not necessarily in a good geographical position at this point to service, support, and provide systems and applications engineering on many of the products they manufacture and export, that there are basic differences between Japanese and U.S. workers that affect the nature of marketing and follow-up and which Japanese companies do not entirely understand. This is where "value added" comes into play. In general, it may be difficult for companies to compete with Japan within, for example, the robotics area if they are merely competing on a product-for-product basis. The best of Japan's robot companies will probably be able to turn out robots at a lower cost than U.S. manufacturers and likely also of very high quality.

Perhaps that is one of the reasons why the United States has viewed licensing agreements between U.S. and Japanese companies as being very positive. U.S. companies, if they selected the right Japanese partner, have found a cost-effective way to get into the market with good products. Japanese companies, if they selected the right U.S. partner, have seen value added to the product in order to make it more attractive in U.S. markets.

Value-added companies are stressing software, computer controls, pre- and postinstallation customer support, an understanding of systems integration, and high-quality industrial design focusing on combining sensing capa-

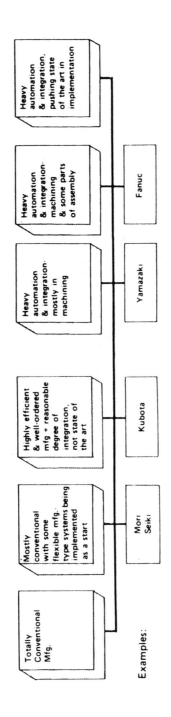

FIGURE 11.1. Manufacturing technology; from conventional to highly automated and integrated.

bilities, controls, and mechanical hardware. They should also be just as adept at incremental improvements within these areas. If one examines the robot industry in the United States from this perspective (adding in the need for strong management and good financial strength or financing capabilities), it is not nearly so difficult to focus in on key companies and to make some manageable sense out of the growing list of companies.

In Japan, out of the more than 150 robot vendors, at this juncture include such companies as Nippon Electric, Fujitsu Ltd., Fanuc, Hitachi, Matsushia, and Yaskawa. This, of course, excludes companies that have particular strengths in certain areas or that are specializing in certain applications or types of robots, where they are likely to continue to be strong.

United States companies will have to be smarter in order to compete effectively with the best of the Japanese companies in the new robotic industries; a selective search within these industries will turn up U.S. value-added companies that are likely to be successful even under these more difficult circumstances.

11.3   Present Trends in Industrial Robotics

Investors should be aware of some important present trends in the robotics industry that are not in the public eye but that will be the key to the way this industry shapes up over the next three years. A shakeout is expected in the robotics industry. There are too many companies worldwide chasing too small a market.

The smaller companies that survive will be characterized principally by product lines that compete through value-added features rather than commodity-type products that compete through price cutting. Among the features that add value are superior computer controllers, software, pre- and postinstallation support, field services, broad-based and more effective distribution networks, overall price/performance, or turnkey systems. The best of these companies will ultimately be characterized by high margins and returns on equity.

By 1983–1985, a continuous flow of these companies is expected into the public marketplace. (Right now many of them are privately held, financed by venture capitalists.) They will not be equally attractive, and investors will want to develop a grasp of how companies fit into these trends before the companies begin to go public.

11.4   Future Trends in Robotics

Vendors and end users will, more and more in the future, be marching to a similar tune. In general, vendors are finally beginning to take their own industry more seriously (one would have thought that with all the capital that they will be pouring into this industry over the next several years this would have already been the case); but based on exhibits we have seen at various shows, many vendors have felt, in the past, more comfortable acquiescing to the "cuteness" inherent in robotics. That is, we have been treated to robots handing out souvenirs (a practice that invariably leads to a crowded booth but not necessarily by the interested potential customer), writing on blackboards, drawing pictures, handling bowling balls, and the like. All these demonstrations had the very valid, dual purposes of (1) attracting attention and (2) demonstrating some of the characteristics of the robot, such as dexterity, flexibility, and agility. The end user was presumably expected to draw the inference that if the robot could, for example, manipulate a placard, sign its name, hand out pens, combs, markers, buttons, bottle caps, and toys, then it could also do some fairly spectacular things on the factory floor. The market share of U.S.-based robot vendors is shown in Figure 11.2.

If robotics is just beginning to bloom, then vision is still in the seedling stage. Although the market for vision systems is still far too thin to generate anything but some fairly broad based conclusions, we believe that the earliest growth of computer-based vision systems will come from independent (i.e., non-robot-related) applications. Nonetheless, it is interesting that venture capitalists, with a timetable for a return on their investments of anywhere from about three to seven years, have done and are doing an increasing amount of investing in the area of computer-based vision start-ups. Perhaps one can use that as somewhat of a guideline for a time frame on when this type of technology would begin to dot the factory floor in numbers at least great enough to generate some sort of measurable return on investment for these venture capitalists. In any event, as with any industry, robotics will also need its early growth period.

It is also generally felt that there are five key elements giving direction to the future of industrial robots. Until recently, progress in one of these areas was largely neutralized by relative stagnation in another. What is now propelling the industry is the fact that each of these five elements, discussed next, is beginning to move in tandem, and highly perceptible strides are being made.

Sensing Devices.   For robots to be truly useful across a wider breadth of markets, they must be able to adjust automatically to production setups. "Blind" robots are adequate in highly organized factory environments where parts positions are changed relatively infrequently or where extensive software packages have already essentially recorded positions of all parts. The majority of American industry is not so ordered. Therefore, for this purpose of opening the market to rapid growth throughout the widest span of industry, robots must be automatically adaptive—capable of recognizing, reorienting, and then manipulating disordered parts. For many assembly and installation procedures, this adaptive ability would be essential.

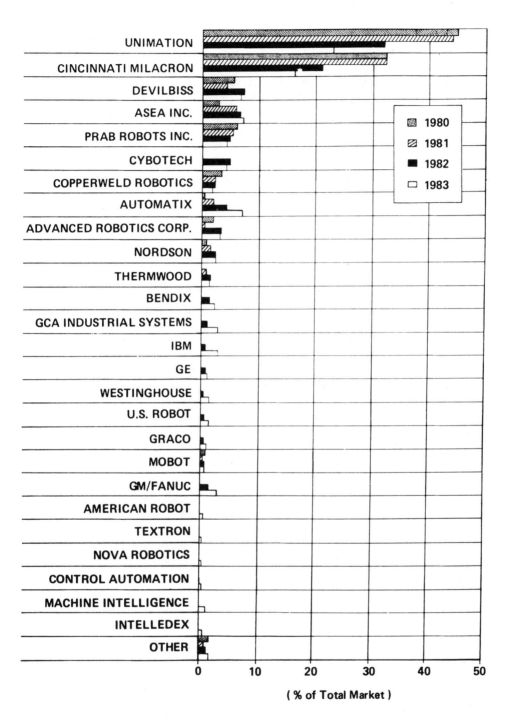

UNIMATION

CINCINNATI MILACRON

DEVILBISS

ASEA INC.

PRAB ROBOTS INC.

CYBOTECH

COPPERWELD ROBOTICS

AUTOMATIX

ADVANCED ROBOTICS CORP.

NORDSON

THERMWOOD

BENDIX

GCA INDUSTRIAL SYSTEMS

IBM

GE

WESTINGHOUSE

U.S. ROBOT

GRACO

MOBOT

GM/FANUC

AMERICAN ROBOT

TEXTRON

NOVA ROBOTICS

CONTROL AUTOMATION

MACHINE INTELLIGENCE

INTELLEDEX

OTHER

1980
1981
1982
1983

0   10   20   30   40   50

(% of Total Market)

* Market share percentages are based on mid-points of company and industry estimates.

Source: Bache Halsey Stuart Shields, Inc.

FIGURE 11.2. Market share of U.S.-based robot vendors for 1980 to 1983.

A considerable amount of research is currently being devoted to the areas of visual and tactile sensing. For example, at NASA's Jet Propulsion Laboratory, researchers are working with "artificial skin" for robots that could adapt to their environment through using the sense of touch to register varying amounts of pressure. A number of visual systems have been developed and are being tested to accomplish pattern recognition in two or three dimensions. Key centers of this type of research include the artificial intelligence center at Stanford University, Draper Laboratories, the National Bureau of Standards, and the University of Rhode Island. Consight-1, developed by General Motors, is a data interpolation system using light and an overhead video camera to direct a robot to grasp, reorient and manipulate parts moving past it on a conveyor. General Electric has been working in the field, as has Auto-Place, whose Opto-Sense purportedly uses a sum-of-the-pixels type of method to accomplish limited visual sensing. Other companies working in the area of visual sensing for robotics and other applications are Solid Photography (SLPH-$3\frac{3}{4}$), Object Recognition Systems, Machine Intelligence, and Unimation. Industry sources are generally agreed that sufficient refinements in terms of processing speed and cost will occur during the decade to allow robots to accomplish complex assembly operations.

Cost. Until fairly recently, industrial robots have been quagmired in a paralysis caused by their high cost relative to other available labor or machinery. Increased labor costs and semiconductor advancements have changed the economics of robots. Furthermore, the range of prices for robots available for a given task has begun to broaden on the lower end. Thermwood's spray painting robot, a continuous path servo-controlled model, at $45,000, is about half the price of competing models. Unimation's PUMA, at about $35,000, is considerably below the prices of other high-technology robots, which range up to approximately $120,000. Additional advancements, increased production volume, and the possible entry of such companies as Texas Instruments or General Electric into the field should result in driving the price of high-technology robots down closer to the area of $10,000, at which point economic feasibility would be well within the reach of even small job shops.

Size. High-technology industrial robots have been largely dominated by machines with work envelopes of up to 1000 ft^3 and load capacity of over 350 lb. To be introduced into factory environments, robots of this size and reach generally require the displacement of existing machinery or even entirely new manufacturing configurations. Certain factory setups are ideally suited to these heavy-duty robotics devices. A smaller robot, however, with a reduced work envelope and payload, would more readily adapt to the vast majority of industry uses. Even in the automotive industry, where nearly half of the heavy-duty robots have been installed, it is estimated that 90% of the parts in the average automobile weigh less than 3.1 lb. Unimation's PUMA 500 and 250 series, with load capacities of 5.0 and 3.5 lb, respectively, are suited not only for manipulation of most automotive parts but for small parts and electronic subsystem assemblies as well. With respective reaches of 34 and 16 in., these servo-controlled robots can be easily integrated into production lines with minimal displacement. Smaller robots such as the PUMA can, of course, also be removed from a production line for maintenance and off-line programming without taking down the whole line.

Systems Integration. Manufacturers are showing increasing interest in the concept of "families of parts" for greater manufacturing efficiency. Briefly, this concept ties together previously independent numerical control machine tools, transfer mechanisms, or parts handling devices, and a coordinating control system to deal with the mid-volume range of production that represents the majority (around 70%) of production. So-called flexible manufacturing systems (FMSs) and manufacturing cells have as their primary aim that of combining good levels of output over a wide range of families of components with the flexibility to change what had previously been achievable only by a sharp reduction in output. Robotics manufacturers are beginning to shift their marketing emphasis from sales of isolated pockets of robots to that of systems design whereby robots are intrinsically linked into the total manufacturing process.

Marketing. The fifth and, in some ways, the most important of the five primary elements directing the future growth of industrial robots is that of aggressive marketing. Now that robots have reached the point of accceptance and have, in fact, become "fashionable," robot manufacturers, the majority of whose backgrounds are in the engineering and technical fields, are facing a market that is ripe for picking. In this regard, manufacturers such as Cincinnati Milacron, with considerable distributive strength through their extensive machine tool sales network, have an edge, as do companies with well-cultivated reputations in the field, such as Unimation. Should Texas Instruments enter, with its substantial marketing muscle, the market will get a boost such as it has not yet experienced.

11.5   Growth in Industrial Robotics[b]

Significant trends are developing in the robotics industry with regard to growth, applications, individual vendors, competition, technologies, users,

[b] Section 11.5 is from Laura Conigliaro, Trends in the Robotics Industry, MS82–122, presented at Robot VI Conference, March 2–4, 1982, Detroit, Mich., and is used courtesy of Laura Conigliaro and Society of Manufacturing Engineers.

importance of the industry, and in other areas. Although most users cited the fact that user interest was growing dramatically and that less looking and more buying was the likely attitude beginning by mid to late 1982, most companies noted that industry growth, though it would be extremely impressive and quite substantial, would not be constant, reflecting economic cycles, the need for end-user and functional diffusion, and user confusion regarding the number of competitors and products. Interestingly, several vendors noted that interest, serious inquiries, and outright purchasing is picking up even now from non-Fortune 500 companies, a fact that could have very positive implications for the overall industry and, particularly, for smaller companies looking for individual niches. Thus, it would appear that user visibility—and not just publicity—is increasing, albeit at a slow pace.

Faster-than-expected developments are cited in the area of assembly robots and growing momentum for arc-welding robots, the latter to be accelerated by likely additions during 1982 of real-time seam tracking devices. Japanese pressure, in particular, will force the exploration of new markets and applications earlier than they might otherwise have been developed.

Strong Japanese competition is expected coming almost entirely for the next few years through licensing agreements between U.S. and Japanese companies and the continuing entry of new vendors into the industry in the United States. The robot industry as a world marketplace inevitably is hampered by the difficulty in providing the necessary "hand-holding" for customers from abroad. This would suggest that vendors here are either going to set up robotics operations abroad or that more licensing agreements will begin to flow from U.S. vendors to foreign companies, and not just vice versa, within the next few years.

Because of the speed with which the robotics industry appears to be moving competitively, several companies formed and run by technologists might be unable to keep pace.

The technologies for robots have primarily emanated out of related industries, such as machine tools or fluid power, but new technologies developed specifically for robots will begin to appear, particularly focusing on user-friendly languages, higher-level generative languages, standard protocol for interfacing and communications devices, easier off-line programming capabilities, controls technology better designed for flexibility, and hardware and software modularity. Modular construction is needed by nearly every user.

The trend toward new materials (high-strength low-alloy steels, plastics, composites) needed for robots implies that more compact, lightweight, and streamlined robots with equivalent strength will become more prevalent. A number of end-users referred to the growing trend away from hydraulically driven robots.

Some users are developing applications engineering expertise in-house, and more users are requiring the robot vendor to take some portion of the turnkey responsibility even though the robot may only represent one aspect of a total system. Several times the intense customer support responsibility of the vendor from before the purchase order has been signed to after the robot has been installed was cited, with reference made to a growing number of vendors who, to one degree or another and in one area or another, would offer some portion of the factory automation systems approach, not just robots.

The current problem is that the present state of the economy, plus the significant degree of hand-holding necessary between vendor and customer, has made customer diffusion all the more difficult, with the difficulty augmented by the growing base of credible competitors in any given application. This has not impacted smaller companies quite so much since they have both smaller bases of operation and also lower projected sales to meet. In fact, in certain instances, we believe that the slowdown has given some necessary breathing time to smaller companies, which may find themselves more able to assimilate developmental, preproduction, or early organizational start-up problems.

A justification for using robots is the potential they offer for enhancing productivity and decreasing cost. Realistically, though, the cost of a robot, as with so many pieces of equipment, is very much a function of a variety of surrounding factors, including previous user experience with robots, the specific type of application, the number of robots being installed, and the alternative method that robots are replacing. Clearly, as the user experience level increases, direct cost efficiencies also increase. It should come as no surprise to anyone that cost curves for many industries in Japan, notably the automotive and steel industries, are considerably lower than our own, with a good deal of modern and flexible automation built into their capital stock. The winners will be those companies that can use methods of lowering themselves on the cost curve and can phase out obsolete production methods. In reverting to embodied technology, it seems fair to say that it is not in stages one or two—finding a better solution to a problem and designing it for the production process where we come up short. Rather, it is more in stage three, the actual embodiment of that technology or its implementation, where the dissimilarities become more apparent.

A further justification for robots is the capability they offer to increase quality and to do it consistently. Inspecting quality into a product through end-of-process inspection is both inadequate and impossible. This means that quality must be manufactured into the product, again an emphasis on process technology.

For the final justification—and this is perhaps the most far-reaching and in some ways the most significant—look to the makeup of the robot itself. As a computer-based product, it both draws from and depends on a database. Thus, it is capable of generating management reports, diagnostics, and, ultimately, of tying into other forms of computerized automation.

Presumably, a goodly number of the large companies that enter the robotics industry or are already in it will have the commitment that is necessary in order to be long-term important contenders. This, plus their unquestioned financial strength, should be sufficient to eliminate the smaller companies—or should it?

It is important to remember that, over the last two years, any number of companies have been formed to try to take advantage of the future potential for robots. Several of them are the results of engineers leaving universities or companies where they were working in the robotics field, having realized that there was a good deal of venture capital financing available for robotics-related start-up companies. In fact, between January of 1980 and the present, over $30 million has been invested by venture capitalists in these companies.

These companies invariably have the commitment and, for some of them, adequate financial backing (relatively speaking) as well. What, then, are they lacking? In many cases, what they are lacking is business acumen. Many of these companies, having been started by talented engineers, are still basically being run by technologists as they go through their early start-up phase.

In the past, it can be shown that a number of high-technology industries began with these types of small companies. The companies often had the luxury of time to travel their own learning curves and to develop slowly before having to meet larger competitors. The result was a whole new series of healthy, strong companies—Intel is a good example in the semiconductor industry—that could compete very effectively as their industry came into its own and as larger companies entered the industry. Small companies in the robotics industry will not have that luxury of time. They already have the specter of large companies facing them. One trend that is clear already is that robot companies managed by technologists will have a rough road ahead of them. To survive and prosper in the industry, companies will need a good product, but they will need more than that as well. As already noted, they will need strength in marketing and field service support, and understanding of the factory floor, on-going R&D, and solid business management skills.

Several of the smaller companies already have a number of these skills and have had important experience in running other enterprises. While their approach may not be unique, the collection of people they have amassed cannot entirely be discounted. Remember that small companies and high-technology companies often have the advantage of attracting certain key people, individuals who desire a more loosely structured environment and who can be enticed by some of the compensation packages that start-up companies can afford to offer or that technology companies are more geared to offer. Several of these companies, despite the smallness of their size, are extremely aggressive and have gotten off to a "running start." Though representing a very tiny group among all the companies in this field, they already show the potential promise of being formidable competitors in their own way and should not be discounted either by end-users or by investors.

Other of these small companies will, by focusing on a particular niche, also be able to survive and be successful in a *relative* sense. With so many ways to cut the end-user market (by application, region, technology, price, etc.), there is ample room for certain of these smaller companies that have properly targeted themselves and are not lured into other markets that will diffuse their focus. Nonetheless, many of their number will clearly never fulfill the promise that they set out to meet and will be largely ignorable over time. At present, all these companies in the aggregate, plus the larger companies as well, are serving a critically useful function, that of focusing attention on the industry and broadening user awareness of it.

11.6   Changes in the Work Force in Industrial Robotics[c]

As robots move into the workplace, the obvious question arises. What happens to the work force? The potential effects of robots and new technology on the work force have only begun to be explored. For many reasons, these effects are difficult to study or predict. Yet this is an area of intense concern to many people.

The few estimates that have been made on the potential displacement of workers vary widely. Many economic factors influence how fast business and industry adopt robots and other new technology.

How fast industry adopts robots is tied into the cost of robots. Some think the price of robots is attractive to industry. Says one analyst, "The robot is more than a machine, it is an irresistible economic force." But for some small- and medium-sized companies, purchasing a robot represents a major investment. Almost all companies considering robots compare the cost of the machine against the labor cost that could be saved. Wage rates, then, affect how valuable robots will be as substitute labor.

The cost of a robot encompasses not only that of the robot itself, but also the expense of fitting it into the existing factory or production line. Sometimes robots may only pay if applied in a whole new facility.

Reasonably priced robots could prove to be attractive to industry and favorably influence their adoption. Likewise, the pressure of foreign competition could push American manufacturers to adopt robotic technology to remain competitive. Finally, the further development and refinement of computer-aided design (CAD) and computer-aided manufacturing (CAM) could be yet another boost to the use of robotic technology in industry.

Even after robotic technology is adopted, it is still difficult to predict its effect on employment. An important issue is whether companies will effectively

[c]Sections 11.6 and 11.7 are from Sar A. Levitan and Clifford M. Johnson, "The Future of Work: Does It Belong to Us or the Robots?" *Occupational Outlook Quarterly*, Fall 1982, Department of Labor, Bureau of Labor Statistics, and are used courtesy of the Department of Labor, Bureau of Labor Statistics.

use robots after they buy them. For example, the Carnegie Mellon University study said one engineer from a company complained that production engineers are so used to thinking in terms of traditional, fixed automation that it is difficult to get them to think in terms of using programmable robots. This is a significant point, since the chief advantage of robots is that they are a flexible technology and generally require only reprogramming, rather than retooling, to be adapted to changes in production. The Carnegie Mellon University study furthermore found that, despite the frequent mention of this flexibility in the trade literature, it was not nearly as important a motivation of their survey respondents as was saving labor costs.

Consequently, generalizations about productivity cannot safely be made solely on the basis of the number of robots in place and the number of companies using robotic and other advanced technology. Given that robots will be used, if robots are not used to their full advantage, they may displace workers without creating productivity gains. On the other hand, they may also have less of an impact on workers.

With these variables in mind, it is easy to see why there is little agreement as to what effects robots and new technology will have on the work force. Let us look at some of the projections that have been made.

As for numbers of workers affected by robots and new technology, estimates range from relatively low to high. The Carnegie Mellon University study offers the most detailed analysis of the number of jobs that potentially could be roboticized. Researchers there conclude that nearly one-half million workers in operative positions in manufacturing in the automotive, electrical equipment, machinery, and fabricated metals industries (the metalworking sectors) could be replaced by nonsensor-based robots. The figure doubles to 1 million people in these same industries if sensor-based robots are considered, says the Carnegie Mellon study. Extrapolating the data for metalworking to other manufacturing sectors, Carnegie Mellon University researchers conclude that nonservo-controlled robots could replace up to 1 million workers, while servo-controlled robots could replace up to 3 million out of a total of roughly 8 million operative workers in manfacturing. According to researchers, the time frame for this displacement is at least 20 years.

The Carnegie Mellon University estimates could be viewed as an upper bound. The study notes that the estimates of the number of jobs that robots could do is not a reliable estimate of how many workers will lose their jobs. But it points out that the figures indicate the magnitude of people who are likely to be affected in a direct way, without distinguishing how many will lose jobs, how many will be transferred, or how many will be retrained.

Harley Shaiken, research fellow at the Massachusetts Institute of Technology and past consultant to the United Auto Workers, offers an estimate at the high end also. He suggests that by 1990 at least 100,000 auto workers alone could be replaced by 32,000 robots. He also suggests that the potential loss of jobs is more serious than these figures indicate, since more advanced robots could displace far more workers in a decade. Moreover, according to Shaiken, "The employment effects are cumulative and have a disproportionate impact on a few key industries."

On the other hand, the Robot Institute of America predicts that only a small segment of the work force will actually lose their jobs as a direct result of robots. According to Peter Blake, the latest study completed for the Institute by the University of Michigan indicates that, by 1990, 440,000 jobs that are now done by humans will be filled by industrial robots. The study suggests that substantial numbers of displaced workers will be retained by their companies and only 22,000 people may actually lose their jobs. The survey concludes that there may even be a net gain in jobs because of robots, since there will be a need for operators and maintenance teams, as well as workers to produce the machines themselves.

Although the number of workers that could be affected by robots is of interest and concern to many persons, it certainly is not the only issue that robots and new technology will create in the work force and in society in general. Equally serious concerns are the following: changes in the nature of the workplace, changes in the composition of the work force, changes in skill requirements of jobs, and the problems of worker retraining and the concomitant shifts necessary in vocational education programs.

An obvious advantage of robots joining the work force is that robots, like other innovative technologies, contribute to the quality of worklife by taking over some dull and hazardous jobs. Richard Beecher, from GM, points out these advantages: "Robots can be applied to those tasks which are least pleasant for our employees and which may require the most extensive safeguards to prevent the possibility of employee injury or exposure to hazardous substances." So, robots can make the workplace safer by doing the worst jobs.

They might also free workers for better jobs. As Beecher says, "The use of robots tends to reduce the number of repetitive, boring jobs and to increase the number of available skilled jobs, particularly in the repair and machine maintenance areas."

Yet even this is not a certain effect. Some see another result of new technology. Russell A. Hedden, president of Cross and Trecker Corp., an industrial equipment manufacturer, says that computer-controlled machine tools can take more and more skill off the shop floor and put it into the computers and machine tools.

Robots have the potential to change the quantity and type of labor demanded by industry. Some analysts suggest that ultimately a work force with a higher skill level will emerge. Speaking at a congressional roundtable on emerging issues, Thomas Weekly, coordinator of the skilled trades department at the United Auto Workers remarked, "Common sense and simple facts show that automation does replace workers, and though a certain amount of new work is shifted to a skilled maintenance group, the long time effect is a reduction in nonskilled labor."

This is a change that can potentially create problems, since it is the unskilled who are least able to find new work if they are put out of work and the most hard-pressed to enter the labor force, especially in times of limited openings for unskilled workers. Some analysts have suggested that this change, if it comes about, will place an increasingly heavy burden on those least able to cope. In addition, it presents a question of who will provide retraining so that these workers can qualify for potentially more plentiful skilled jobs.

But there is also a possible good side to this change in the proportion of skilled to nonskilled jobs in the economy. Perhaps larger numbers of more interesting jobs will open up to which certain workers cannot now aspire. For example, a skilled trade worker such as an electrician may be put in charge of maintaining and supervising a robot welder. Hence, the worker's responsibilities are expanded.

Another effect robots and new technology could have on the work force is a trend toward shiftwork, especially in industries not now using shift scheduling. Robotic specialists and maintenance personnel may be expected to be available around the clock if that is when the robots work.

11.7   New Occupations in the Robotic Age

New jobs and new occupations might emerge as a result of the new technology. Analysts generally agree that new jobs will be created with the coming of robots; but, as with potential worker displacement, estimates on what type of jobs and how many will be created is a subject of debate.

The introduction of robots into the workplace will create jobs in the programming, maintenance, and operation of robots. According to Peter Blake, "These will definitely be jobs not previously found on shop floors." Also, the increased use of robotic technology is creating demand for robotics engineers and manufacturing engineers experienced in robotic applications. Engineers with related specialties will also be in demand for research and development of robots by users as well as by robot manufacturers. In addition, successful use of robots in industry will require computer specialists to develop the proper software for applications.

According to Don Smith, Director of the Industrial Development Division of the Institute of Science and Technology at the University of Michigan, robotic maintenance workers will represent the largest number of new jobs. According to the Carnegie Mellon University study, different skills are required for operation and maintenance. While maintenance duties require more extensive retraining, operation or monitoring duties are essentially unskilled and are more easily picked up. The levels and skills required also differ according to the type and complexity of robots.

For persons involved in the operation of robots, elaborate retraining is unnecessary. Almost anyone can be taught to operate a robot with very minimal training. Typically, experienced maintenance people require a week of formal training to adapt their skills to routine servicing and monitoring of robots. Frequently, the robot manufacturers or vendors provide the training programs that give participants the necessary skills to maintain and operate robots. Workers are also often given on-the-job training of several weeks following the training programs. Some of the large robot users, such as General Motors, employ their own training specialists within the plant to ensure that their workers have the skills to maintain robots and other types of electronic systems.

More extensive instruction is necessary to learn the skills associated with robot maintenance. Skill requirements in the primary stages of robot maintenance include knowledge of hydraulic valves, basic electricity, numerical controls, and other areas of general plant and machine maintenance. Although some of the courses needed to learn to maintain a robot are very technical, they do not require skills very different from those involved in maintaining other pieces of complex machinery, according to the Carnegie Mellon University study. Researchers of the study concluded that anyone who has experience in machine maintenance can be easily retrained to maintain robots.

Besides maintenance workers, programmers are also needed for developing software for robotic applications, and it is expected that computer software specialists for manufacturing systems will be much in demand. "It is in the software instructions that robots represent a state-of-the art development," says David Babson, publisher of the *Babson Weekly Staff Letter*.

In the engineering profession, a new manufacturing engineer will be in demand—the robotics specialist. This engineer must be capable of planning the robot installation of programmable machines, including robots, and the specifications for their capability and work environment and designing the production line around the capabilities of programmable automation.

Besides working on applications for industrial users, manufacturing engineers will be especially in demand at computer companies that enter the robot manufacturing market. According to David Babson, most computer firms, especially those in the mainframe business, have years of experience in software development, but they lack an intimate knowledge of machinery and how factory procedures differ from industry to industry.

New employment opportunities will be found not only in businesses that purchase and use robots but also in robot manufacturing companies. The robot manufacturing industry will employ between 60,000 and 100,000 persons by 1990, based on estimates done for the Society of Manufacturing Engineers. If the robot business expands, businesses that supply products to robot manufacturers will also grow. Robot manufacturers will employ persons chiefly in three areas: machine construction, computers, and software. Engineers will be needed, as well, to design and direct the assembly of robots. In addition, personnel will be needed to market robots and provide customer services.

11.8   Education in Industrial Robots[d]

11.8.1   Robotics at Vincennes University

The Electronics Technology Department at Vincennes University will begin a two-year program in Robotics in the fall of 1983. The program will start after three years of preparation and will provide the students with an A.S. degree in robotics technology. This program has recently been approved and partially funded by the Indiana State Board of Vocational and Technical Education, with additional funding requested from the state legislature. In the tradition of Vincennes University, this high-technology program will feature state-of-the-art equipment with hands-on approach lab time providing students with the skill, training, and education needed to service robots and perform application engineering to robots in industry.

The concept of robots to the general public in the past has been the R_2-D_2 of "Star Wars" and other manlike creatures; however, with the publicity in magazines, newspapers, and television today, more exposure has been given to robots in industry, which show the device to be a large armlike mechanism assisted by either hydraulics, pneumatics, or dc servomotors controlled by microprocessors or computer-based systems. The basic difference of the machines as compared to other numerical-controlled or computer-controlled machines is that the robots are not a dedicated system. They can be used for one application, be programmed to perform other applications, or even perform several applications within the same work week. The flexibility of the programmable robot is truly assisting in the "New" or "Second Industrial Revolution." These robots today basically are used in industry for welding, painting, material handling, palletizing, assembly, and for jobs too dangerous or health hazardous such as die casting and other jobs where material must be handled in poor environments.

The "factories of the future," which are in the planning stages today, will largely use robots for batch production, repetitive jobs, and hazardous jobs. These industrial robots can duplicate the human's best work on his best day, hour after hour, day after day, with more speed and better consistency in accuracy. These robots are the answer to the industrial problem of quality assurance and productivity. What will become of the factory worker? How many jobs will be displaced? A few years ago we heard the same questions about computers. Now with the microprocessors in industry, homes, and businesses, more jobs have been created than displaced by computers. So it will be with robotics; more jobs will be created than displaced by this new industry. As with any increase in the level of technology, the untrained or the unskilled must prepare for a higher-technology society.

---

[d]The material in Section 11.8.1 is used courtesy of Richard Dean Eavey, Professor, Electronics Dept., Vincennes University, Vincennes, Ind.

It has been forecasted that companies such as Ford Manufacturing will install 5000 robots in the next 10 years, while GM predicts 10,000 to 15,000 robots will be installed. Jeep is now installing about 30 robots, and many other companies worldwide are using robots for all types of applications. The need for robotic technicians is estimated to be approximately 300,000 in the next 10 years. The Society of Manufacturing Engineers, Robotics International Division, states that we need programs in the universities to prepare technicians to maintain these high technology machines.

At Vincennes University, the robotics program will provide very specialized training in robotics as an option program to the electronics technology program. The new program will use 12 Rhino robot trainers tied to 12 Apple computers for the introduction courses. Advanced courses will use industrial robots such as the Cincinnati Milacron T^3. Several other units are also being purchased. Other devices such as Allen Bradley and Modicom programmable controllers and automated conveyor systems will be used in the program.

Students will take electronics courses such as DC/AC, transistor circuit design, digital electronics, microprocessor software, microprocessor hardware, industrial electronics control, and robotics courses: introduction to robotics, hydraulics and pneumatics, robotic interfacing, robotic servicing, and robotic application engineering.

The program offered the first course in January 1983 as a pilot course and the complete program began in August 1983. Enrollment was limited to 48 freshmen.

Besides at Vincennes University, robotic servicing and programming are taught at Piedment, Greenwood, South Carolina, Henry Ford Institute, Michigan, McCoomb Junior College, Detroit, and other community colleges.

Senior colleges teaching robotics courses include the following:

University of California, Davis, Davis, California

Carnegie-Mellon University, Pittsburgh, Pennsylvania

Georgia Institute of Technology, Atlanta, Georgia

University of New Mexico, Albuquerque, New Mexico

New York Institute of Technology, Old Westbury, New York

Purdue University, West Lafayette, Indiana

Rutgers, The State University of New Jersey, Piscataway, New Jersey

Tennessee Technological University, Cokeville, Tennessee

University of Texas at Arlington, Arlington, Texas

University of Toronto, Toronto, Ontario, Canada

Tufts University, Medford, Massachusetts

Western Michigan University, Kalamazoo, Michigan

Worcester Polytechnic Institute, Worcester, Massachusetts

Oregon State University, Corvalis, Oregon

Polytechnic Institute of New York, Brooklyn, New York

Some junior and senior colleges only teach one robotics course and others provide degree programs.

11.8.2   Robotic Technology Option

A robotics technology option at Vincennes University is presented. This curriculum will prepare graduates to enter the robotics field as service technicians, engineering assistants, and manufacturing and industrial maintenance technicians. Students gain additional laboratory experiences in electronic fundamentals, communication techniques, digital systems, and microprocessors.

FIRST YEAR

		Course	Sem I	Sem II
TEL	110	DC and AC Circuit Analysis	5	
TEL	125	Transistor Circuit Design	4	
SMT	106	Applied Mathematics II	3	
HEW	101	English Composition I	3	
		Physical Education Activity	1	
TRO	150	Introduction to Robotics		3
SPT	160	Applied Physics of Hydraulics and Pneumatics		3
TEL	170	Digital Systems and Computer Circuitry		5
SMT	107	Applied Mathematics III		3
SNT	150	Computer-Aided Circuit Analysis		3
			16	17

SECOND YEAR

		Course	Sem I	Sem II
TRO	200	Robotic Interfacing	3	
TRO	210	Servicing Automated Manufacturing Systems	2	
TEL	230	Microprocessor Software and Interfacing	3	
		Social Science Elective	3	
		Approved Humanities Elective	3	
		Physical Education Activity	1	
TRO	250	Robotic Applications		5
TEL	260	Microprocessor Hardware		3
TEL	275	Industrial Electronics Control		5
		Social Science Elective		3
			15	16

New course descriptions in robotic technology at Vincennes University follow:

TRO 150 Introduction to Robotics. *3 hours [Semester II]:* An introduction to industrial robotic applications in industry, classification of robotic systems, and product familiarization. Basic application engineering projects will be approached. Lab will consist of programming robots using pick and place and teach modes by using operator keyboards. Mechanical systems within the robot will be explored, with emphasis on maintenance. (3 hours lecture, 5 hours lab weekly)

TRO 200 Robotics Interfacing. *3 hours [Semester I]:* Introduction is based on hardware and software necessary for connecting a microprocessor to a robotic arm, and to interface the system with other machines such as welders, painters, conveyors, programmable controllers, etc. Additional emphasis will focus on actuators and end-effectors for robotic arms. (3 hours lecture, 5 hours lab weekly)

TRO 210 Servicing Automated Manufacturing Systems. *2 hours [Semester I]:* This course provides instruction in servicing robotic systems and conveyor systems. Emphasis is on mechanical servicing, hydraulic servicing, and associated electronic circuits. (2 hours lecture, 3 hours lab weekly)

TRO 250 Robotics Applications. *5 hours [Semester II]:* This special applications course will climax the study of robotics in industry. A class project will center around the manufacturing of a product, completely automated by robotics and conveyor systems with programmable controllers and microprocessors synchronizing the manufacturing process. Students will set up the system as a semester project with experience with a product that will be palletized, welded, painted, and assembled. Necessary hardware and software will be developed for production, with emphasis on servicing, troubleshooting, and applications engineering. (5 hours lecture, 7 hours lab weekly)

SPT 160 Applied Physics of Hydraulics and Pneumatics. *3 hours [Semester II]:* A laboratory science course applied to the theory and operation of hydraulic and pneumatic systems, logic flow diagrams for hydraulics and pneumatics. Specific emphasis on feedback loops controlled by logic electromechanical devices. (2 hours lecture, 2 hours lab weekly)

11.8.3 Electromechanical Technology Robotic Core

A suggested associate degree curriculum for electromechanical technology core with specialty options follows:

CORE PROGRAM

FIRST SEMESTER

Course	Title	Theory	Lab	Credits
		[Hours/week]		
PR1101	Physics—Fluids	2	3	3
AT1102	Algebra and Trigonometry	3	—	3
GE1103	Humanities	3	—	3
DM1104	Mechanics	2	2	3
GE1105	Technical Communications	1	0	1
DC1106	DC Theory and Circuits	2	3	4
		13	8	17

SECOND SEMESTER

PL1201	Physics—Optics	2	2	3
TM1202	Technical Maths	3	0	3
GE1203	Social Science	2	0	2
ED1204	Semiconductors	3	2	4
DP1205	Computer Literacy	2	2	3
AC1206	AC Theory and Circuits	2	2	3
		14	8	18

THIRD SEMESTER

AE2101	Analog Electronics	2	2	3
DE2102	Digital Electronics	2	2	3
FB2103	Instrumentation and Control	2	2	3
MP2104	Microprocessors	2	2	3
CM2105	Communications Circuits	2	2	3
TD2106	Technical Drawing	2	0	2
		12	10	17

SPECIALTY OPTIONS

FOURTH SEMESTER

Course	Title	Theory	Lab	Credits
		[Hours/week]		
[A]	ROBOTICS TECHNOLOGY			
IS2201	Industrial Safety	2	0	2
PA2202	Principles of Automation	2	2	3
RC2203	Robot Construction	2	2	3
RP2204	Robot Programming	2	4	4
RS2205	System Servicing	1	4	3
CD2206	Contemporary Developments	2	0	2
		11	12	17

Other high technology options could include:

[B] Laser and Fiber Optics Technology
[C] Computer Systems Technology
[D] CNC Systems Technology
[E] Instrumentation Technology

In the real world an element of politics influences the choice of course titles. Creative titling can help meet the general education requirements of associate degree programs and can help to ease transfer problems. This explains why some of the following course descriptions are not more clearly identified by the course titles. Course descriptions follow:

PF1101 Physics—Fluids: Universal concepts of differential forces, flow rate, real opposition, energy storage, time constants, impedances, and resonance. Laboratory sessions will concentrate on study of pneumatic and hydraulic devices for sensing, control, and actuation.

AT1102 Algebra and Trigonometry: Review of the algebraic operations, exponents, and radicals, simple equations, graphs, systems of equations, quadratic equations. Logarithms, solutions of triangles, complex numbers, trigonometric functions and equations.

GE1103 Humanities: Suitable course to meet formal AAS degree requirements selected from general education catalog.

DM1104 Mechanics—Materials, Drives, and Mechanisms: An introduction to contemporary engineering materials. Properties of stress, strain, elasticity, tensile strength, compressibility, shear resistance, coefficients of friction and thermal expansion. Basic mechanical components: gears, pulleys, levers, bearings, springs, and applications to commonly encountered mechanisms.

GE1105 Technical Communications: Approaches to report writing, report contents and organization, grammer and readability. Instruction incorporates actual reports written by students for current lab courses. Methods of collecting information, basic library science. Preparation and delivery of technical presentations. Interview techniques and preparation of resumés.

DC1106 DC Theory and Circuits: Basic electric units, resistance, capacitance and inductance, series and parallel circuits, superposition and Thevenin's theorems. Small dc motors and generators.

PL1202 Physics II—Optics: Emphasizes geometrical and physical optics covering wave motion, interference, diffraction, and polarization. Introduces lasers and fiberoptics technology. Reviews heat units, conductivity, convection, and radiation. Reviews sound waves, resonance, and harmonics.

TM1202 Technical Maths: Reinforces selected arithmetic, geometric, algebraic, and trigonometric functions of particular importance to electromechanical technicians. Ratios, percentages, Pythagoras theorems, vector addition, linear and quadratic equations, sine and cosine values, complex numbers. Concepts of differential and integral calculus as rates of change, Laplace transforms, graphical solutions.

GE1203 Social Science: Appropriate course selected from General Education catalog to meet formal requirements of AAS degree curriculum.

ED1204 Semiconductor Theory, Devices, and Circuits: Brief introduction to atomic structure, insulators, conductors, and semiconductors. The semiconductor junction, diode, and transistor characteristics. Diode rectifiers, clamps, and limiters. Common-emitter, common-base, and common-collector transistor connections. Unijunctions, FETS, triacs, and ACR devices and characteristics.

DP1205 Computer Literacy: Introduces student to major components of computer systems: computers, video terminals, disk stores, and printers. Fundamentals of BASIC language, structured programming for real-time applications.

AC1206 AC Theory and Circuits: Sinusoidal wave forms; instantaneous, peak, and rms values, behavior, and resistors, capacitors, and inductors as ac. Series and parallel circuits, reactance, impedance, resonance. Power in ac circuits. Small ac motors. Rectifiers and filters.

AC2101 Analog Electronics: Amplifiers, with special emphasis on operational amplifiers. Thyristor circuits. Oscillators, phase-locked loops, filters and attenuators. Analog-to-digital converters. Introduction to feedback systems.

DE2101 Digital Electronics: Numbering systems, gating circuits, combinational logic. Multivibrators, pulse shaping, clocks and timing circuits, registers, sequential logic. TTL, CMOS, ECL, and other "families." Digital-to-analog converters. Sampling and multiplexing.

FB2103 Instrumentation and Control: Transducers, bridge techniques, telemetry principles. Switched and continuous control systems. Servomotors, tachometers, servo pots, synchros, stepping motors, and encoders. Servomechanism principles.

MP2104 Microprocessors and Interfacing: Architecture, address decoding, bus systems, standard interfaces. Assembly language, data manipulation, transfer of control, digital I/O, analog I/O, interrupts, and polling techniques. Applications examples. Development aids.

CM2105 Communications Circuits: An overview course introducing principles of AM and FM systems. Multiplexing, PCM, FSK, ASK. Communication protocols, buffers, baud rate, standard codes, ASCII, ISO, etc.

TD2106 Technical Drawing: Interpretation of engineering drawings, importance of correct dimensioning and tolerancing. Preparation and interpretation of circuit diagrams with standard symbols. Assembly drawings, production flow charts and layouts.

IS2201 Industrial Safety: Protection of personnel, product, and production systems. OSHA regulations. Safe plant layout, fail-safe controls, safe operating procedures, development of safety training programs and emergency plans.

PA2202 Principles of Production Automation: Levels of automation; justifications of automation, elimination of hazardous tasks, increasing throughput, improving product quality. Return on investment. Sociological implications, selling robots to the work force, retraining displaced workers. Introduction to time and motion studies and plant layout for efficient automation.

RC2203 Robot Construction: Cartesian, cylindrical, spherical, and revolute systems. Arm geometry; drive systems, electric, hydraulic, pneumatic, open and closed loop. Dynamic performance. End-effectors. Teach pendants, computer interfacing, voice- and touch-sensitive command systems. Vision systems. Safety interlocks.

RP2204 Robot Programming: Hardware restrictions, initializing open-loop systems. Levels of programming, machine, point-to-point, primitive motion, structured programming. Comparison of current robot languages. User friendly features, menu driven, and conversational programs.

RS2205 System Servicing Techniques: Safety of personnel and protection of system while troubleshooting. Troubleshooting aids, logical fault isolation, diagnostic software. Device identification, locating alternative devices, device replacement techniques. Recalibration and burn-in procedures. Preventive maintenance.

CD2206 Contemporary Developments: A tutorial course introducing new devices, systems, applications, and regulations as they become available in the marketplace.

A mechanical robotic core could include courses in:

1. Microprocessor systems in factory production
2. Fluid power

3. Robotic transducers
4. Applied mechanisms
5. Mechanical systems of robots
6. Programming of robots
7. Servicing of robots
8. Practical mechanics for robots
9. Manufacturing systems
10. Computer-aided design drafting
11. Mechanical design for electronics production
12. Hydraulics and pneumatic systems
13. Safety for robotic systems

Each community or senior college must adapt its own programming needs.

11.8.4 Electromechanical Robotic Program for High School Students

The following draft outlines the subject areas and *some* recommended equipments to assist with an electromechanical program for high technology and robotics.

Prerequisites to high school level include the following:

1. Mathematics
2. Physics
3. Chemistry

Program of study includes:

1. Electrical fundamentals
2. Mechanism and kinematics
3. Technical drawing
4. Basic electronics and transducers
5. Fluid power fundamentals
6. Digital electronics
7. Switching logic
8. Safety
9. Introduction to numerical control (NC)
10. Microprocessors and programming
11. Open and closed loop control
12. Microcomputers and real-time control
13. Introduction to three-term (PID) process control
14. Robotics I

15. Work safety
16. Industrial numerical control (NC)
17. Robotics II
18. Industrial psychology
19. Robotics III
20. Industrial robotic systems

Four or more robotic manufacturing firms produce educational robots for use in robot technology programs in the high schools, community, and senior colleges, and technical societies. These include the following.

1. Feedback, Inc., is a specialist source of instructional hardware and support services for technical and engineering education. The company provides laboratory equipment for high-technology programs, including communications, control, computers, and robotics. Feedback, Inc., is located in Berkeley Heights, N.J.

2. The Heath Company has introduced a multifunction robot and an accompanying robotics education course on how to constuct it. Priced at approximately $1500 in kit form or $2495 assembled, the Heath robot, named HERO 1, is both intelligent and mobile. It is a self-contained, electromechanical robot controlled by its own onboard, programmable computer and carries electronic sensors to detect light, sound, motion, and obstructions in its path. It can be programmed to pick up small objects with its arm, speak complete words and sentences with its voice synthesizer, travel over predetermined courses, and repeat specific functions on a predetermined schedule. It is equipped with a rechargeable power supply and can move freely without external guidance. HERO 1 stands 20 in. tall and weighs 39 lb. It can be programmed with either its onboard keyboard or an external training pendant supplied in the package. Programming can be stored on any standard audio cassette tape recorder and reloaded for future use. The Heath Company is located at Benton Harbor, Michigan.

3. Microbot, Inc., is also a specialist in instructional hardware in robotics. They manufacture the Minimover and Teachmover robots, which sell for under $2700. They also have the Alpha robot for industrial applications. They are located at Menlo Park, California.

4. The Sandhu Machine Design Company of Champaign, Illinois, produces the Rhino[R] XR-1 for use in community colleges for robotic technology programs; it costs about $2400.

It is important to consider the software options required for programming. In the future, more robots will be available to enhance a proper educational program in robotics.

11.9   Industrial Robots and
Factory Automation[e]

11.9.1   Industry Overview

The industry has developed more slowly than most observers anticipated, with initial uses confined to the handling of hazardous materials or the performance of duties in difficult environments. In the last several years the robot industry has exhibited impressive growth characteristics, primarily because American industry has attempted to increase its productivity in order to compete with foreign competitors. Nevertheless, most robots used today are still hazardous environment machines (HEMs), and only recently have businesses attempted to replace semiskilled assembly line workers with robots. Hostile environment machines in industry are used for spot welding, spray painting, investment castings, and other tasks in unpleasant environments.

At present, the major end-market user of robots is the automotive industry, where high volume and competitive concerns make the application of robotics an attractive choice. An estimated 40% of all industrial robots go to the automotive industry, followed by heavy machinery and aerospace industries, each representing approximately 10% of robot industry sales.

The economics of robot use are compelling, with frequently cited total cost of the machines at about $5.00/hour, compared with almost triple that for UAW wage rates. Nevertheless, the true advantages of robotic systems lie in more than the simple tradeoff between manpower and machinery. Robotics is thought to result in greater productivity for users through reduced scrappage, better quality control, and better materials control. Most studies show that the majority of users of industrial robots view the most important functional areas of robotics to be assembly operations, spray painting, spot welding, machine loading, material handling, and die casting.

In coming years, HEM could exhibit slightly slower growth than the market for assembly robots because the market for HEM has been established for some time. Specifically, Frost and Sullivan project the industry increasing from $26 million in 1977 to $438 million in 1985. While not as optimistic a projection, the International Resource Development Corporation projects an average annual growth rate of 15% in real terms through 1984. Similar projections also appear in recent articles in *Fortune Magazine* and *Business Week*.

---

[e]The material in Sections 11.9 and 11.10 is used courtesy of Wertheimer and Co., Inc., © 1981.

11.9.2   The Machines

There are currently two basic types of robots, those that are mathematically based (which necessarily require digital computers as their base) and those that are not. Cincinnati Milacron, Unimation, and U.S. Robotics are some of the major American companies that offer the former so-called "smart" technology.

Industrial robots may further be characterized by their weight-carrying capability, be it light, medium, or heavy. The light-load robots are often known as assembly robots and may have either electric motor drives or pneumatic drives. Medium-load robots are either electrically or hydraulically driven. Heavy-load robots are hydraulically driven.

In the future, robot control may develop from using a combination of sensory inputs and mathematically based controls. These techniques will enable a robot to identify objects and then carry out an instruction regarding what to do with such objects from large databases, which are created by CAD systems. With parts fully defined by the CAD systems, sensors would be able to recognize and locate the objects, and the robot could act on these objects according to some predescribed schedule, producing the finished results. The teaching of each portion of the system would be done by computers, obviating the need for long setup times for production facilities, and allowing users the full benefits that robots can offer—flexible automation. The future possibilities of the technology will depend heavily on the development of the semiconductor industry and new software methods from the artificial intelligence community at universities around the world. Other areas that will advance commercially available robot characteristics are electric and hydraulic actuators and controls for robots and the incorporation of materials with higher strength-to-weight characteristics.

11.9.3   The Competition

In many respects the robot industry is in its infancy, and the survival of current competitors will depend more on the type and quality of their product than on the length of time that they have been in business. As other companies enter the robot market and offer more advanced products, the established robot manufacturers may or may not be able to stay current with any such design innovations so offered.

The U.S. industrial robot market is currently dominated by Unimation and Cincinnati Milacron. While neither company details sales for robotic systems, it is estimated that the robot sales of the two companies are of about equal size and between them constitute the major portion of the domestic robot industry. Cincinnati Milacron, entering the robot business only several years ago, has apparently adopted the strategy of competing mainly in the HEM market, with its T-3 model marketed almost entirely to welding applications. Cincinnati Milacron appears to have had very good success with this specific end market.

Unimation has been in the robot business for almost 20 years and offers a broader product line than Cincinnati Milacron. Its Unimate model has competed in the more established large machine market, while its PUMA model competes in the smaller, assembly-type market. Unimation's end markets as well as functional applications are thought to be broader than those of Cincinnati Milacron.

Smaller companies also continue to enter and perform in the robot industry:

1. American Robot Company, a North Carolina-based company, has spent several years developing a low-cost robot that incorporates less expensive materials in its construction.
2. Automatix was founded by Vic Scheinman (a former employee of Unimation and designer of the PUMA model) and Phil Villers (a co-founder of Computervision) to market robot systems. Automatix currently plans to initially sell robot systems manufactured by others on a turnkey basis, and only later commence manufacturing its own robot systems.
3. Auto-Place, recently acquired by Copperweld, which in turn is primarily owned by Imetal, a French concern, builds small pick-and-place robots of about 5-lb payload capacity with high quality, but pneumatically driven.
4. Cybotech, a joint venture between Ransburg and Renault, is thought to be initially selling a large robot, but planned to introduce a computer-based assembly robot in 1982.
5. General Electric has introduced its Allegro assembly robots, based upon the Pragma robots manufactured by DEA, an Italian corporation.
6. Hitachi and Fujitsu Fanuc, both Japan-based companies, are independently producing sophisticated computer-based robots.
7. Kulicke & Soffa is thought to be entering the field of industrial robots, although it has indicated it is between 12 and 24 months away from its initial placement of a robot.
8. Moog, a leading supplier of servovalves to the aerospace industry, derives about 5% of its sales from valve sales to robotics. However, its estimated 80% market share is in jeopardy due to increased purchases by Unimation from Kawasaki and Cybotech from Renault.
9. Prab Conveyor has acquired AMF's Versatran line, a product line of large non-computer-based robots that has been produced since 1969.
10. Robomation Corporation, an entity with sales estimated at approximately $500,000 per year, markets small, modular robots that are nonmathematically based.

11. Seiko Instruments Company, a division of Dari-Seikosha Ltd. of Japan, is a consumer products company that has been producing high-quality small robots for several years, although its robots are pneumatically driven.

12. United States Robots, a manufacturer of smart assembly robots, was founded by two former Unimation employees. Its first product, The Maker, is now on the market, and other high-technology offerings are expected to follow.

13. IBM has introduced several robots; one that has received a great deal of attention is the RS-1 cartesian robot system for precision assembly work.

At the present time, there are about 70 robot manufacturers in the United States. New entrants include Texas Instruments, which already uses a smart assembly-type robot for the testing of calculators. Texas Instruments has had small assembly robots for its internal use for several years, but has not emphasized the commercialization of its robots.

11.9.4 Automated Material Handling Systems

More efficient handling of materials was a constant obsession of factory management during the 1960s and early 1970s. Financing charges were low and the prime motivation was to reduce the labor content of material handling processes. The major concern of the 1980s will be the cost of financing work-in-process inventory, which is estimated to be on the factory floor 80% to 90% of the time and on the machines only 10% to 20% of the time. Industrial robotics must also be considered in automated material handling systems.

Material handling systems are designed to increase the flow of products in the plant. These systems should be controlled by the central host computer in the factory, rather than the current method of using isolated lift trucks controlled manually by the operator. These automated conveyor systems further reduce the labor content in the material handling process. High stackers, above 30 ft, are manufactured principally by the Kenway Division of Eaton and cost about $100,000. They can replace 15,000 lift trucks, with the 20-year life of a high stacker equivalent to about four generations of lift trucks. Therefore, the equivalent truck cost would be about $183,000, justifying the investment without including the reduced labor content as well as the direct tie into the computer on the factory floor.

A second advance in the factory of the future is the automatic guided vehicle system (AGVS), a component in Cincinnati Milacron's Variable Mission Manufacturing System. Advances in AGVS systems have come about from breakthroughs in their control systems, advances in the controls of automatic storage and retrieval systems (AS/RS), and the conversion to computer numerical control for many machine tools, the latter allowing again the tie-in to the central host. The principal manufacturer of these automatic systems is the Kenway Division of Eaton, accounting for about 5% of Eaton's sales. Automatic storage and retrieval systems in smaller sizes than Kenway (15 to 30 ft height capability) is a fast-growing market, presently dominated by Raymond Corporation, manufacturer of small, narrow-aisle lift trucks. Use of such trucks is also growing relative to general lift trucks as they are needed to load automatic material handling systems.

The other major areas of automated material handling systems are conveying equipment, including drag chain conveyors, tow line conveyors, and the more exotic "power and free" conveying systems used in nonsynchronized lines, as well as warehousing conveyors, used in automated warehousing systems. The last two areas are dominated by S. I. Handling systems, which should continue to participate in the growth of these segments.

11.10   Industrial Robots: The Delphi Survey

The Delphi Survey reflects collected opinions from managers and engineers in industrial robots. (See reference to Smith and Wilson in the Bibliography.)

Q: By what year will 20% of mass production companies be using dedicated, automatically assembled assembly machines?

A: Median—1987.

Q: By what year will practical robots be available which have some form of vision systems?

A: Median—1985.

Q: When will small batch production be performed by programmable robots for small batch assembly in the batch sizes set out below?

A: 200, 1985; 500, 1990; 1000, 1992.

Q: By the year indicated, what will be the proportion of dedicated machinery to robots?

A: 1985, 85 to 15; 1990, 70 to 30; 1995, 60 to 40.

Comment: From a negligible base, industrial robots are estimated to increase their share of the enormous assembly market. Within just 10 years, robots are estimated to be able to perform almost all the assembly work of humans, growing at a phenomenal rate relative to the population of dedicated machinery in existence.

11.11   Epilogue

We hope our readers have read some chapters that interested them. The field of robotics will some day find that it has no enemies and no secrets. The field will change in the next 10 to 25 years in the right direction. The basic concepts in this book will remain the same. Costs will rise, people will not change. The authors have tried to stick to basics that do not change and make this a text that everyone can use and enjoy. Our ideas will change and so will industrial robotics.

In about 20 years, with advances made in computer technology, with trillion bits of memory per second, and sophisticated programming, it will be possible for robots to think and act as humans. Robots will work with other robots and solve intelligence problems that even people cannot do at this time in 1983. Programming of multiple tasks will then be feasible.

In about 10 years from now, robots with fast memory will be used in the offices of industrial manufacturing and the home to service workers and people of leisure.

Robots will eventually take a place in our technological revolution to change our personal destiny. Modular approaches in robotics will assist the growth with standardization in the robotic elements.

The factory of the future can use high-powered carbon dioxide lasers in industrial robotic systems to move parts among several manufacturing processes. In the future, the laser beam can be time-shared to perform cutting, welding, and other tasks. Flexible machining systems can be used to move parts from one machining operation to another. Fiberoptic devices will also possibly perform in the integrated manufacturing cell of the future.

Places of learning such as the University of Rhode Island, the Artificial Intelligence Laboratories at Stanford University, MIT, Carnegie Mellon, and other senior colleges are developing machine vision systems that will aid industry and government agencies in future growth.

The University of Rhode Island has created software systems that permit a robot arm to do a particular task such as bin-picking through machine vision. Such systems will allow robotic growth in spite of the fact that robots do dangerous, uncomfortable, fatiguing, or just repetitive and unpleasant tasks. Growth will mean that in the future robots will help humanity and change our style of living.

11.12   Review Questions

1. Discuss present trends in economics for industrial robotics.
2. Is there any end to smaller companies in industrial robotics?
3. Discuss factory automation in industrial robotics.

4. Discuss the new occupations in industrial robotics.

5. Discuss the courses required in electronic, mechanical, and electromechanical systems to be able to do robotic servicing.

6. Discuss the growth of industrial robotics now, 10 years from now, and 50 years from now.

7. Discuss the meaning of HEM.

8. Discuss the impact of industrial robotics on American productivity.

9. Will the United States ever compete with Japan in industrial robotics?

11.13   Bibliography

"Assembly Technology—# 1 Growth Opportunity for Robots," *Assembly Engineering,* Vol. 26, no. 6, June, 1983.

Laura Conigliaro, "Trends in the Robot Industry (Revisited): Where Are We Now? Robots 7 Conference, Apr. 1983, Chicago, Illinois.

Engineering Education, Engineering College Research and Graduate Study, vol. 79, no. 6, published yearly by American Society of Engineering Education, Washington, D.C..

Industrial Robots International, vol. 3, no. 21, Nov. 8, 1982.

Sar A. Levitan and Clifford M. Johnson, "The Future of Work: Does It Belong to Us or to the Robots?, *Monthly Labor Review,* U.S. Department of Labor, Bureau of Labor Statistics, Sept. 1982.

Manufacturing Technology Horizons, Mar./Apr. 1983.

Gail M. Martin, "Industrial Robots Join the Work Force," *Occupational Outlook Quarterly,* U.S. Department of Labor, Bureau of Labor Statistics, Fall 1982.

Mitchell I. Quain and James B. Townsend, "Factory Automation, Investment Recommendations," Wertheimer and Co., Inc., 200 Park Ave., New York, New York, © 1981, Wertheimer and Co., Inc.

Tom Parrett, "The Rise of The Robot," *Science Digest,* vol. 91, no. 4, Apr. 1983, pp. 68–75 and 107.

"Robotic Newsletter," nos. 1 through 11, Institutional Research published by the Prudential Bache, periodically through the year and written by Laura Conigliaro.

Donald N. Smith and Richard C. Wilson, *Industrial Robots, a Delphi Forecast of Markets and Technology,* 1st edition, 1st printing, Society of Manufacturing Engineers/The University of Michigan, Ann Arbor, 1982.

Special Robotics Section, *Omni Magazine,* Apr. 1983, pp. 6–154.

Appendix A

A Glossary of Commonly Used Robot Terms

Actuator: A motor or transducer that converts electrical, hydraulic, or pneumatic energy to effect motion of the robot.

Adaptable: Capable of making self-directed corrections. In a robot, this is often accomplished with visual, force, or tactile sensors.

Adaptive control: A control method in which control parameters are continuously and automatically adjusted in response to measured process variables to achieve better performance.

Analog control: Control involving analog signal processing devices (electronic, hydraulic, pneumatic, etc.).

Arm: An interconnected set of links and powered joints comprising a manipulator that supports or moves a wrist and hand or end-effector.

Axes: The directions of movements, for example X, Y, Z, as in numerical machine tool control.

Bang-bang control: A binary control system that rapidly changes from one mode or state to the other (in motion systems, this applies to direction only).

Cell: A manufacturing unit consisting of two or more work stations and the material transport mechanisms and storage buffers that interconnect them.

Courtesy of the National Bureau of Standards and prepared for the U.S. Air Force ICAM Program. Appears in *Iron Age Magazine*, Mar. 19, 1982. Also courtesy of *Iron Age*.

Center: A manufacturing unit consisting of two or more cells and the materials transport and storage buffers that interconnect them.

Closed-loop control: Control achieved by feedback, i.e., by measuring the degree to which actual system response conforms to desired system response and utilizing the difference to drive the system into conformance.

Cognitive intelligence: Ability to plan and establish goals, and to model the environment based upon sensory input.

Computed path control: A control scheme wherein the path of the manipulator end point is computed to achieve a desired result in conformance to a given criterion, such as an acceleration limit, a minimum time, etc.

Computer control: Control involving one or more electronic digital computers.

Continuous-path control: A control scheme whereby the inputs or commands specify every point along a desired path of motion.

Cycle time: The period of time from starting one machine operation to starting another (in a pattern of continuous repetition).

Cylindrical coordinate robot: A robot whose manipulator arm degrees of freedom are defined primarily by cylindrical coordinates.

Dead zone: A range within which a nonzero input causes no output.

Digital: Away from the base, toward the end-effector of the arm.

Digital control: Control involving digital logic devices, which may or may not be complete digital computers.

Drift: The tendency of a system's response to gradually move away from the desired response.

Duty cycle: The fraction of time during which a device or system will be active or at full power.

Dynamic accuracy: Degree of conformance to the true value when relevant variables are changing with time.

Elbow: The joint that connects a robot's upper arm and forearm.

End-effector: An actuator, gripper, or mechanical device attached to the wrist of a manipulator by which objects can be grasped or acted upon.

End point control: Any control scheme in which only the motion of the manipulator end point may be commanded, and the computer can command the actuators at the various degrees of freedom to achieve the desired result.

End-point rigidity: The resistance of the hand, tool, or end point of a manipulator arm to motion under applied force.

Error signal: The difference between desired response and actual response.

Factory: A manufacturing unit consisting of two or more centers and the materials transport, storage buffers, and communications that interconnect them.

Fixed stop robot: A robot with stop point control but no trajectory control; i.e., each of its axes has a fixed limit at each end of its stroke and cannot stop except at one or the other of these limits. Such a robot with N degrees of freedom can therefore stop at no more than two locations (where location includes position and orientation). Often very good repeatability can be obtained with a fixed stop robot.

Flexible: (1) Pliable or capable of bending. In robot mechanisms, this may be due to joints, links, or transmission elements. Flexibility allows the end point of the robot to sag or deflect under load and to vibrate as a result of acceleration or deceleration. (2) Multipurpose; adaptable; capable of being redirected, restrained or used for new purposes. Refers to the reprogrammability or multitask capability of robots.

Gripper: A device by which a robot may grasp and hold external objects.

Group technology: A system for coding parts based on similarities in geometrical shape or other characteristics of parts. The grouping of parts into families based on similarities in their production so that parts of a particular family could be processed together.

Intelligent robot: A robot that can be programmed to make performance choices contingent on sensory inputs.

Interface: A shared boundary. An interface might be a mechanical or electrical connection between two devices; it might be a portion of computer storage accessed by two or more programs; or it might be a device for communication to or from a human operator.

Joint: A rotational or translational degree of freedom in a manipulator system.

Limited degree-of-freedom robot: A robot able to position and orient its end-effector in fewer than six degrees of freedom.

Manipulator: A mechanism usually consisting of a series of segments, jointed or sliding relative to one another, for the purpose of grasping and moving objects usually in several degrees of freedom. It may be remotely controlled by a computer or by a human.

Microcomputer: A computer that uses a microprocessor as its basic element.

Microprocessor: The principal processing element of a microcomputer made as a single integrated circuit.

Modular: Made up of subunits that can be combined in various ways. In robots, a robot constructed from a number of interchangeable subunits each of which can be one of a range of sizes or have one of several possible motion styles (prismatic, cylindrical, etc.) and number of axes.

Multiprocessor: A computer or network of computers that can execute several programs concurrently under integrated control.

Multiprocessor control: A control scheme that employs more than one central processing unit in simultaneous parallel computation.

Open-loop robot: A robot that incorporates no feedback, i.e., no means of comparing actual output to commanded input of position or rate.

Passive accommodation: Compliant behavior of a robot's end point in response to forces exerted on it. No sensors, controls, or actuators are involved. The remote center compliance provides this in a coordinate system acting at the tip of a gripped part.

Pick-and-place robot: A simple robot, often with only two or three degrees of freedom, that transfers items from place to place by means of point-to-point moves. Little or no trajectory control is available. Often referred to as a "bank-bank" robot.

Pitch: The angular rotation of a moving body about an axis perpendicular to its direction of motion and in the same plane as its top side.

Point-to-point control: A control scheme whereby the inputs or commands specify only a limited number of points along a desired path of motion. The control system determines the intervening path segments.

Programmable: Capable of being instructed to operate in a specified manner or of accepting set points or other commands from a remote source.

Programmable controller: A controller whose algorithm for computing control outputs is programmable.

Programmable manipulator: A device that is capable of manipulating objects by executing a stored program resident in its memory.

Proximity sensor: A device that senses that an object is only a short distance (e.g., a few inches or feet) away, and/or measures how far away it is. Proximity sensors work on the principles of triangulation of reflected light, lapsed time for reflected sound, or intensity induced eddy currents, magnetic fields, back pressure from air jets, and others.

Rate control: Control system in which the input is the desired velocity of the controlled object.

Record-playback robot: A manipulator for which the critical points along desired trajectories are stored in sequence by recording the actual values of the joint position encoders of the robot as it is moved under operator control. To perform the task, these points are played back to the robot servo system.

Remote center compliance [RCC]: A compliant device used to interface a robot or other mechanical workhead to its tool or working medium. The RCC allows a gripped part to rotate about its tip or to translate without rotating when pushed laterally at its tip. The RCC thus provides general lateral and rotational "float" and greatly eases robot or other mechanical assembly in the presence of error in parts, jigs, pallets, and robots. It is especially useful in performing very close clearance or interference insertions.

Robot: A reprogrammable multifunctional manipulator designed to move material, parts, tools, or specialized devices through variable programmed motions for the performance of a variety of tasks. [Editorial Note: This RIA robot definition has been accepted worldwide at the recent 11th International Symposium on Industrial Robots (ISIR) in Tokyo.]

Robot programming language: A computer language especially designed for writing programs for controlling robots.

Sensor: A transducer whose input is a physical phenomenon and whose output is a quantitative measure of the physical phenomenon.

Sensory control: Control of a robot based on sensor readings. Several types can be employed: (1) Sensors used in threshold tests to terminate robot activity or branch to other activity. (2) Sensors used in a continuous way to guide or direct changes in robot motions. (3) Sensors used to monitor robot progress and to check for task completion or unsafe conditions. (4) Sensors used to retrospectively update robot motion plans prior to the next cycle.

Sensory-controlled robot: A robot whose program sequence can be modified as a function of information sensed from its environment. Robot can be servoed or non-servoed. (Also see *Intelligent Robot.*)

Sensory hierarchy: A relationship of sensory processing elements whereby the results of lower-level elements are utilized as inputs by higher-level elements.

Sensory intelligence: The ability to understand the input signals related to the working area.

Servo-controlled robot: A robot driven by servomechanisms, i.e., motors whose driving signal is a function of the difference between commanded position and/or rate and measured actual position and/or rate. Such a robot is capable of stopping at or moving through a practically unlimited number of points in executing a programmed trajectory.

Slew rate: The maximum rate at which a system can follow a commanded motion.

Speed–payload tradeoff: The relationship between corresponding values of maximum speed and payload with which an operation can be accomplished to some criterion of satisfaction, and with all other factors remaining the same.

Spherical coordinate robot: A robot whose manipulator arm degrees of freedom are defined primarily by spherical coordinates.

Supervisory control: A control scheme whereby a person or computer monitors and intermittently reprograms, sets subgoals, or adjusts control parameters of a lower-level automatic controller while the lower-level controller performs the control task continuously in real time.

Tactile sensor: A transducer that is sensitive to touch.

Teach: To program a manipulator arm by guiding it through a series of points or in a motion pattern that is recorded for subsequent automatic action by the manipulator.

Teaching interface: The machanisms or devices by which a human operator teaches a machine.

Teleoperator: A device having sensors and actuators for mobility and/or manipulation, remotely controlled by a human operator. A teleoperator allows an operator to extend his or her sensory-motor function to remote or hazardous environments.

Transducer: A device that converts one form of energy into another.

Twist: Rotational displacement around a reference line; same as roll.

Upper arm: That portion of a jointed arm that is connected to the shoulder.

Work station: A manufacturing unit consisting of one or more numerically controlled machine tools serviced by a robot.

Working envelope: The set of points representing the maximum extent or reach of the robot hand or working tool in all directions.

Working range: All positions within the working envelope. The range of any variable within which the system normally operates.

Wrist: A set of rotary joints between the arm and hand that allows the hand to be oriented to the workpiece.

Yaw: The angular displacement of a moving body about an axis that is perpendicular to the line of motion and to the top side of the body.

Appendix B

A Glossary
of Commonly Used Terms
in Robotics and Computers

Access time: The time interval between the instant at which information is: (1) Called for from storage and the instant at which delivery is completed, i.e., the read time. (2) Ready for storage and the instant at which storage is complete, i.e, the write time.

Accuracy: A measurement of a deviation from a straight line or a particular taught point in space. Accuracy deviations are attributed to calculation errors, arm geometry errors, and poor READY location alignment.

Adaptive: The ability of the robot to "learn," modify its control system, and respond to a changing environment.

Addend: The number or digital quantity to be added to another (augend) to produce a result (sum).

Address: A label, name, or number that designates a register, a memory location, or a device.

Air motor: A device that converts compressed air into rotary mechanical force and motion. An air-servo-motor is one that is controlled by a servo mechanism.

Algorithm: A fixed step-by-step procedure for accomplishing a given result.

Alphanumeric or alphameric: Characters that are either letters of the alphabet, numerals, or special symbols.

Courtesy of International Robomation/Intelligence.

Analog: The representation of numerical quantities by means of physical variables; i.e., translation, rotation, voltage, or resistance.

Analog-to-digital [A/D] converter: A device that changes physical motion or electrical voltage into digital factors.

AND: A logical operator that has the property such that if X and Y are two logic variables, then the function "X AND Y" is defined by the following table:

X	Y	X AND Y
0	0	0
0	1	0
1	0	0
1	1	1

The AND operator is usually represented in electrical notation by a centered dot ".", and in FORTRAN programming notation by an asterisk "*" within a Boolean expression.

Anthropomorphic: Resembling human shape or characteristics. This term is used to describe the ability of the robot arm to move in a fashion similar to the human arm.

AROM: Alterable ROM. A ROM which can be changed or initialized by the user.

Articulated arm: A robot arm constructed to simulate the human arm, consisting of a series of rotary motions and joints, each powered by an air motor.

Artificial intelligence: The capability of a computer to perform functions that are normally attributed to human intelligence, such as learning, adapting, recognizing, classifying, reasoning, self-correction and improvement.

Asynchronous: Not occurring at the same time.

Binary picture: Result of a *thresholding operation* performed on an image.

Blob: A cluster of adjacent one's in a binary image.

BPS: Bits per second.

Branch: (synonymous with Jump) An instruction that when executed may cause the arithmetic and control unit to obtain the next instruction from some location other than the next sequential location. It is one of two types, conditional or unconditional.

Breakpoint: A point in the program where it may be interrupted by external intervention.

BSC: Binary Synchronous Communication. A communication line discipline used for controlling the exchange of digital data between computers and/or terminals across telephone lines.

Bus: A channel along which data can be sent.

Byte: A sequence of binary digits usually operated upon as a unit.

Calling sequence: A basic set of instructions used to begin, initialize or transfer control to and return from a subroutine.

Cartesian coordinates: A set of three numbers defining the location of a point within a recti-linear coordinate system consisting of three perpendicular axes (X, Y, Z).

Character: (synonymous with Item) One of a set of elements which may be arranged in ordered groups to express information. Each character has two forms: (1) A man-intelligible form, the graphic, including the decimal digits 0–9, the letters A–Z, punctuation marks, and other formatting and control symbols. (2) A computer-intelligible form, the code, consisting of a group of binary digits (bits).

Chip: An integrated circuit.

Clamp: Function of the pneumatic hand that controls grasping and releasing of an object.

Clear: To replace information in a storage unit by zero (or blank, in some machines).

Closed loop: A control system that uses feedback. An open loop control system is one that does not use feedback.

Coding: To prepare a set of computer instructions required to perform a given action or solve a given problem.

Coding, symbolic: Coding in which the instructions are written in nonmachine language.

Compile: To prepare with a compiler an object language program from a symbolic language program by substituting machine-operation codes for the symbolic operation codes.

Compiler: A processor program for a scientific procedural programming system.

Compiler language: A computer language, more powerful than an assembly language, which instructs a compiler in translating a source language into a machine language. The machine-language result (object) from the compiler is a translated and expanded version of the original.

Complex sensors: Vision, sonar, and tactile sensors that will enable a robot to interact with the work environment.

Computer: A device capable of accepting information, applying prescribed processes to the information, and supplying the results of these processes.

Conditional jump: The jump is subject to the result of a comparison made during the program.

Contents: The information in a storage location.

Continuous path: A trajectory control system that enables the robot arm to move at a constant tip velocity through a series of predefined locations. A rounding effect of the path is required as the tip attempts to pass through these locations.

Contouring: Controlling the path of the robot arm between successive positions or points in space.

Control, open-loop: An operation where the computer applies control action directly to the process without manual intervention.

Control unit: That portion of a computer that directs the automatic operation of the computer, interprets computer instructions, and initiates the proper signals to the other computer circuits to execute instructions.

Convolution: A generally applicable local operation defined by the formula:

$$q[i,j] = \frac{1}{c_0} \sum_{n=-k}^{+k} \sum_{m=-k}^{+k} c_k \cdot f[i,j], \quad k = 1,2, \ldots$$

Depending on the chosen coefficients c_k, the convolution operator may be used to perform such different operations as noise filtering, edge extraction, contour following, and others.

Counter: A device or location that can be set to an initial number and increased or decreased by an arbitrary number.

CPS: Characters per second.

Cycle: A sequence of operations that is repeated regularly. The time it takes for one such sequence to occur.

Data: A collection of facts, numeric and alphabetical characters, etc., which is processed or produced by computer.

Data link: Equipment that permits the transmission of information in data format.

Data processing: A procedure for collecting data and producing a specific result.

Debugging: The process of determining the correctness of a computer routine, locating any errors, and correcting them. Also, the detection and correction of malfunctions in the computer itself.

Deductive capability: Conclusions drawn from knowledge, rules and general principles.

Diagnostic routine: A test program used to detect and identify hardware malfunctions in the computer and its associated I/O equipment.

Differential positioning: The position difference obtained by providing pulses of compressed air to the air motor in opposite directions, resulting in more accurate positioning.

Digit: One of the n symbols of integral value ranging from 0 to $n-1$ inclusive in a scale of numbering of base n, e.g., one of the ten decimal digits—0, 1, 2, 3, 4, 5, 6, 7, 8, 9.

Digital: A description of any data that is expressed in numerical format.

Digital control: The use of a digital computer to perform processing and control tasks in a manner that is more accurate and less expensive than an analog control system.

Digital image: Digital representation of an image, given by a discrete function $f(i,j)$, $0 = i = M$ and $0 = j = N$.

Digital image analysis: A multistage process that leads to the "understanding" of a digital image, the recognition of certain objects, or the recognition of certain attributes in given objects in the image. Stages of the image analysis process may be image digitizing, image preprocessing, feature extraction, pattern recognition.

Digital-to-analog [D/A] converter: A device that transforms digital data into analog data.

Disk memory: A nonprogrammable, bulk-storage, random-access memory consisting of a magnetizable coating on one or both sides of a rotating thin circular plate.

Dithering: A technique used to overcome break-away friction by periodic excitation of the part to be moved.

DNC: Direct Numerical Control. A system in which a digital computer is directly connected to one or more numerically controlled machine tools and controls the machining operations.

Documentation: The group of techniques necessarily used to organize, present, and communicate recorded specialized knowledge.

Dump: A small program that outputs the contents of memory onto hard copy which may be listings, tape, or punched cards.

Duplex circuit: A telegraph circuit permitting simultaneous two-way operation.

Duplex, full: Method of operation of a communication circuit where each end can simultaneously transmit and receive.

Duplex, half: Permits one-direction, electrical communication between stations. Technical arrangements may permit operation in either direction but not simultaneously. Therefore, this term is qualified by one of the following suffixes: S/O for send only; R/O for receive only; S/R for send or receive.

Edit: A computer mode in the RCL system that allows creation or alteration of a program. This mode is available with the robot arm passive.

Encoder: A transducer used to convert position data into electrical signals. The robot system uses an incremental optical encoder to provide position feedback for each joint. Velocity data is computed from the encoder signals and used as an additional feedback signal to assure servo stability.

Engineering units: Units of measure as applied to a process variable, e.g., psi, degrees F, etc.

EPROM: Eraseable Programmable Read Only Memory. A non-volatile memory used to store VAL.

Error: The difference in value between actual response and desired response in the performance of a controlled machine, system or process.

Executive control program: A main system program designed to establish priorities and to process and control other programs.

Feature: Certain characteristics of an image such as edges, contours, silhouettes, transitions from black to white, or vice versa.

Feature extraction: The process of generating a two-dimension characteristic function defined as a $M \times N$ boolean matrix f_F such that:

$$f_F[i,j] = \begin{array}{l} 1 \text{ if } f[i,j] \text{ belongs to the image} \\ 0 \text{ if } f[i,j] \text{ does not belong to the feature} \end{array}$$

Feedback: The signal or data fed back to a commanding unit from a controlled machine or process to denote its response to the command signal. The signal representing the difference between actual response and desired response that is used by the commanding unit to improve performance of the controlled machine or process.

File maintenance: The processing of a master file required to handle the nonperiodic changes in it. For example, changes in number of dependents in a payroll file, the addition of new checking accounts in a bank.

Filter: Device to suppress interference that would appear as noise.

Flag: A bit(s) used to store one bit of information. A flag has two stable states and is the software analogy of a flip-flop.

Flip-flop: A bistable device. A device capable of assuming two stable states. A bistable device which may assume a given stable state depending upon the pulse history of one or more input points and having one or more output points. The device is capable of storing a bit of information, controlling gates, etc. A toggle.

Flow chart: A graphical representation of a sequence of operations, using symbols to represent the operations, such as: Compute, substitute, compare, GO TO, IF, read, write, etc. Flow charts of different levels of generality can be drawn for a given problem solution. Terms used are system flow chart, program flow chart, coding level flow chart, etc.

Format: The predetermined arrangement of characters, fields, lines, page numbers, punctuation marks, etc. Refers to input, output and files.

FORTRAN: "Formula Translator." The language for a scientific procedural programming system.

FORTRAN compiler: A processor program for FORTRAN.

Fusing: An operation that causes separate blobs to fuse into one larger blob.

Gap: An interval of space or time associated with an area of data processing activity (record) to indicate or signal the end of that record.

Garbage: Unwanted and meaningless information in memory.

Gear: The gear with the greater number of teeth in a gear pair (the one with the lesser number of teeth is called the pinion.)

Geometric processing: The process of taking measurements that are characteristic for the geometry of certain objects in an image. Examples are: area (size), coordinates of center of gravity, orientation, perimeter, number and location of holes.

Global operation: Transformaton of the gray scale value of the picture elements according to the gray scale values of all elements of the picture. Examples are Fourier transform, correlation, etc.

Gradient: A vector indicating the change of gray scale values in a certain neighborhood of a pixel. The gradient can be obtained by applying a difference operation on the neighborhood. Often only the magnitude is calculated as a means to obtain the contour of an object.

Graphic system: A system that collects, uses and presents information in pictorial form.

Gripper: A "hand" of a robot that picks up, holds, and releases the part or object being handled. Sometimes referred to as a manipulator or compliance.

Hard copy: Any form of computer-produced printed document. Also, sometimes punched cards or paper tape.

Hardware: The mechanical, magnetic, electrical, and electronic devices of which a computer is built.

Head: A device, usually a small electromagnet on a storage medium such as magnetic tape or a magnetic drum, that reads, records, or erases information on that medium. The block assembly and perforating or reading fingers used for punching or reading holes in paper tape.

Hexadecimal: A notation of numbers in the base 16.

High-level language: A simplified computer programming language that uses English-like statements for instructions and is oriented to the program to be solved or the procedure to be used.

Histogram: Relative frequency of the gray scale value distribution in a digital image. The histogram of an image provides very important basic information about the image.

IC: Integrated Circuit. A solid-state microcircuit contained entirely within a chip of semiconductor material, generally silicon.

Image: A photographic picture, e.g., as being picked up by a TV camera. Mathematically, an image can be described by a function of 2 variables $f(x,y)$, usually defined over a rectangular region. x and y are the region coordinates, and $f(x,y)$ represents the gray scale value of point (x,y) in the region.

Image enhancement: The process of enhancing the quality of the appearance of an image. Image enhancement operations may be noise filtering, contrast sharpening, edge enhancement, etc.

Image preprocessing: A computational step prior to the feature extraction step in an image analysis procedure. Preprocessing may serve the purpose of enhancing the features to be extracted or adjusting the image in other ways to certain conditions set forth by the subsequent feature extraction procedure.

Image processor: Selects and interprets data to determine an object's position, location, shape, and size.

Index: An integer used to specify the location of information within a table or program.

Index register: A memory device containing an index.

Initialize: A program or hardware circuit that will return a program, a system, or a hardware device to an original state.

Input: The date supplied to a computer for processing. The device employed to accomplish this transfer of data.

Instruction: A set of bits that will cause a computer to perform certain prescribed operations. A computer instruction consists of: (1) An operation code which specified the operation(s) to be performed. (2) One or more operands (or addresses or operands in memory). (3) One or more modifiers (or addresses or modifiers) used to modify the operand or its addressee.

Instruction register: A counter that indicates the instruction currently being executed.

Interface: A concept involving the specification of the inter-connection between two equipments having different functions.

Interrupt: A break in the normal flow of a system or program occurring in such a way that the flow can be resumed from that point at a later time. Interrupts are initiated by signals of two types: (1) Signals originating within the computer system to synchronize the operation of the computer system with the outside world (e.g., an operator or a physical process). (2) Signals originating exterior to the computer systems to synchronize the operation of the computer system with the outside world (e.g., an operator or a physical process).

Joint: A single degree of arm rotation. There are up to six joints in a robot arm.

Joint interpolated motion: A method of coordinating the movements of the joints such that all joints arrive at the desired location simultaneously. This method of servo control produces a predictable path regardless of speed and results in the fastest cycle time for a particular move.

K: An abbreviation for 1000 taken from the metric term kilo. Generally used as a measurement of memory capacity. A memory with a capacity of 1K words actually has 1024 words (2^{10}).

Knowledge engineering: The use of artificial intelligence techniques and a base of information or knowledge (facts, rules and procedures) about a specific activity to control systems automatically. This type of system is called a "knowledge-based system."

Label: An ordered set of characters used to symbolically identify an instruction, a program, a quantity, or a data area.

Labeling: The process of assigning different numbers to the picture elements of different blobs in a binary image.

Language: A defined group of representative characters or symbols, combined with specific rules necessary for their interpretation. The rules enable an assembler or compiler to translate the characters into forms (such as digits) meaningful to a machine, a system, or a process.

LED: Light Emitting Diode. A solid-state device used for signal indication on the manual control and the I/O module.

LED display: An alphanumeric display consisting of an array of LEDs.

Left justified: A field of numbers (decimal, binary, etc.) that exists in a memory cell, location or register, possessing no zeroes to its left. For example,

---

3   7   2   ,   2   4   0

Library programs: A software collection of standard routines and subroutines by which problems and parts or problems may be solved on a given computer.

Line printer: A printing device that can print an entire line of characters all at once.

Linkage: A means of communicating information from one routine to another.

Loader: A program that operates on input devices to transfer information from off-line memory to on-line memory.

Local operation: Transformers of the gray scale value of the picture elements according to the gray scale values of the element itself and its neighbors in a given neighborhood. Examples are: gradient, sharpening, smoothing (noise filtering), edge extraction, etc.

Location: A storage position in memory uniquely specified by an address.

Log: A record of values and/or action for a given function.

Loop: The repeated execution of a series of instructions for a fixed number of times.

LPM: Lines per minute.

LSI: Large Scale Integration. High density integration of circuts for complex logic functions. LSI circuts can range up to several thousand logic elements on a one-tenth square inch silicon chip.

Machine language: A language written in a series of bits which are understandable by, and therefore instruct, a computer. The "first level" computer language, as compared to a "second level" assembly language or a "third level" compiler language.

Macro: A source language instruction from which many machine-language instructions can be generated (see compiler language).

Magnetic disk storage: A storage device or system consisting of magnetically coated metal disks.

Magnetic tape storage: A storage system that uses magnetic spots, representing bits, on coated plastic tape.

Mainframe computer: The principal computer in a system of computers.

Manual control: A device containing controls that manipulate the robot arm and allow for the recording of locations and program motion instructions.

Memory: A device or media used to store information in a form that can be understood by the computer hardware.

Memory, bulk: Any nonprogrammable large memory, i.e., drum, disk.

Memory cycle time: The minimum time between two successive data accesses from a memory.

Memory, random access: A memory whose information media are organized into discrete locations, sectors, etc., each uniquely identified by an address. Data may be obtained from such memory by specifying the data address(es) to the memory, e.g., core, drum, disk, cards.

Message: A group of words, variable in length, transporting an item of information.

Micro array computer: A special-purpose multiprocessor system designed for high speed calculations with arrays of data.

Microprocessor: A single integrated circuit containing most of the elements of a computer.

Microsecond: One-millionth of a second.

Millisecond: One-thousandth of a second.

Modem: A contraction of modulator-demodulator. The term may be used with two different meanings: (1) The modulator and the demodulator of a modem are associated at the same end of a circuit. (2) The modulator and the demodulator of a modem are associated at the opposite ends of a circuit to form a channel.

Monitor: An operating programming system that provides a uniform method for handling the real-time aspects of program timing, such as scheduling and basic input/output functions.

Multiplex: The transmission of multiple data bits through a single transmission line by means of a "sharing" technique.

Multiprogamming: A technique for handling numerous routines or programs seemingly simultaneously by overlapping or interleaving their execution, that is, by permitting more than one program to time-share machine components.

Network: The interconnection of a number of devices by data communication facilities. "Local networking" is the communications network internal to a robot. "Global networking" is the ability to provide communications connections outside of the robot's internal system.

Noise: An extraneous signal in an electrical circuit capable of interfering with the desired signal. Loosely, any disturbance tending to interfere with the normal operation of a device or system.

Numerical control: The control of machine tools by mechanical devices.

Octal: Pertaining to the number base 8. Pertaining to any set of exactly eight characteristics.

Offline programming: Computer program development on a system separate from the computer onboard a robot. *On-line programming* is a computer program development on the system included in a robot.

Offset: The count value output from an A/D converter resulting from a zero input analog voltage. Used to correct subsequent nonzero measurements.

Operating system: A group of programming systems operating under control of a data-processing monitor program.

Operation, parallel: Operates on all bits of a word simultaneously.

Operation, serial: The flow of information through a computer in time sequence, usually by bit but sometimes by characters.

Output: Information transferred from the internal storage of a computer to output devices or external storage.

Parameter: In a subroutine, a quantity that may be given different values when the subroutine is used in different main routines or in different parts of one main routine, but that usually remains unchanged throughout any one such use. In a generator, a quantity used to specify input/output devices, to designate subroutines to be included, or otherwise to describe the desired routine to be generated.

Parity bit: A binary digit appended to an array of bits to make the sum of all the bits always odd or always even.

Parity check: A check that tests whether the number of ones (or zeros) in an array of binary digits is odd or even.

Patch: A section of coding inserted into a routine to correct a mistake or alter the routine. Explicitly, transferring control from a routine to a section of coding and back again.

Peripheral: Input/output equipment used to make hard copies or to read in data from hard copies (typer, punch, tape, reader, line printer, etc.). Paper tape is considered hard copy for this definition.

Picture element: An element of the matrix of gray scale values $f(i,j)$.

Pixel: Short for *picture element*.

Point operation: Transformation of the gray scale value of the picture elements according to a given function. Examples are thresholding, contrast enhancement, etc.

Program: A plan for the solution of a problem. A complete program includes plans for the transcription of data, coding for the computer, and plans for the absorption of the results into the system. The list of coded instructions is called a routine. To plan a computation or process from the asking of a question to the delivery of the results, including the integration of the operation into an existing system. Thus programming consists of planning and coding, including numerical analysis, systems analysis, specification of printing formats, and any other functions necessary to the integration of a computer in a system.

PROM: Programmable ROM. A ROM that can be changed or initialized once by the user.

Pseudo [OP] instruction: A symbolic representation of information to a compiler or interpreter. A group of characters having the same general form as a computer instruction, but never executed by the computer as an actual instruction.

Queue: Waiting lines resulting from temporary delays in providing service.

RAM: Random Access Memory. The PUMA system volatile memory used for storage of user programs and locations.

Range: A characterization of a variable or function. All the values that a function may possess.

Read: To copy, usually from one form of storage to another, particularly from external or secondary storage to internal storage. To sense the meaning of arrangements of hardware. To sense the presence of information in a recording medium.

Reader: A device capable of sensing information stored in an off-line memory media (cards, paper tape, magnetic tape) and generating equivalent information in an on-line memory device (register, memory locations).

Real time: Pertaining to the actual time during which a physical process transpires. Pertaining to the performance of a computation during the actual time that the related physical process transpires, in order that results of the computation can be used in guiding the physical process.

Record: A collection of fields. The information relating to one area of activity in a data processing activity, i.e., all information on one inventory item. Sometimes called *item.*

Register: A memory device capable of containing one or more computer bits or words. A register has zero memory latency time and neglible memory access time.

Repeatability: As opposed to accuracy, a measurement of the deviation between a taught location point and the played-back location. Under identical conditions of load and velocity, this deviation will be finer than accuracy tolerance. The closeness of agreement among the number of consecutive movements made by the robot arm to a specific point.

Reset: To return a register or storage location to zero or to a specified initial condition.

Resolution: The resolution of a digital image is given by the values M and N and the number of bits by which the gray scale values of $f(i,j)$, are represented.

Right justified: A field of numbers (decimal, binary, etc.) which exists in a memory cell, location, or register possessing no significant zeros to its right is considered to be right justified. 0 0 0 1 2 0 0 0 0 0 is considered to be a seven-digit field, right justified. 0 0 0 0 0 1 2 0 0 is a two-digit field, not right justified.

Robomation: A contraction of the words "robot" and "automation" meaning the use of robots to control and operate equipment or machines automatically.

Robot: A robot is a reprogrammable multifunctional manipulator designed to move material, parts, tools, or specialized devices, through variable programmed motions for the performance of a variety of tasks.

ROM: Read Only Memory. A digital memory containing a fixed pattern of bits generally unalterable by the user.

Routine: A series of computer instructions that performs a specific, limited task.

RS-232C: Standard computer interface data link used by CRT and TTY terminals.

Scan: To examine signals or data point by point in logical sequence.

Scanner, analog input: A device that will, upon command, connect a specified sensor to measuring equipment and cause the generation of a digit count value which can be read by the computer.

Segmented program: A program that has been divided into parts (segments) in such a manner that: (1) Each segment is self-contained. (2) Interchange of information between segments is by means of data tables in known memory locations. (3) Each segment contains instructions to cause (or request a Monitor to cause) the transfer to the next segment.

Sensor: A transducer or other device whose input is a quantitative measure of some external physical phenomenon and whose output can be read by a computer.

Serial interface: A method of data transmission that permits transmitting a single bit at a time through a single line. Used where high speed input is not necessary. Requires only one wire.

Serial operation: A type of information transfer performed by a digital computer in which all bits of a word are handled sequentially.

Servo mechanism: A control system for the robot in which the computer issues commands, the air motor drives the arm, and a sensor measures the motion and signals the amount of the motion back to the computer. This process is continued until the arm is repositioned to the point requested.

Set point: The required or ideal value of a controlled variable, usually preset in the computer or system controller by an operator.

Shift: To move information serially right or left in a register(s) of a computer. Information shifted out of a register may be lost, or it may be reentered at the other end of the register.

Shrinking: An operation that causes a large block to shrink into a smaller blob.

Sign: The symbol or bit that distinguishes positive from negative numbers.

Signal processing: Complex analysis of wave forms to extract information.

Significant digit: A digit that contributes to the precision of a numeral. The number of significant digits is counted beginning with the digit contributing the most value, called the "most significant digit," and ending with the one contributing the least value, called the "least significant digit."

Simulator: A device or computer program that performs simulation.

Software: The programs or routines, and supporting documentation, that instruct the operations of a computer.

Source language: The symbolic language comprised of statements and formulas used to specify computer processing. It is translated into object language by an assembler or compiler, and is more powerful than an assembly language in that it translates one statement into many items.

Statement: In computer programming, a meaningful expression or generalized instruction in a source language.

Subroutine: A series of computer instructions to perform a specific task for many other routines. It is distinguishable from a main routine in that it requires, as one of its parameters, a location specifying where to return to the main program after its function has been accomplished.

Symbol table: A table of labels and their corresponding numeric values.

Symbolic coding: Broadly, any coding system in which symbols other than actual machine operations and address are used.

Syntax: The structure of expressions in a language. The rules governing the structure of a language.

System: A collection of parts or devices that forms and operates as an organized whole through some form of regulated interaction.

Table: A collection of data, each item being uniquely identified either by some label or by its relative position.

Table look-up.   A procedure for obtaining the function value corresponding to an argument from a table of function values.

Thresholding operation: The process of generating a $M \times N$ boolean matrix f_B such that:

$$f_B[i,j] = \begin{array}{l} 1 \text{ if } f[i,j] \leq \text{threshold} \\ 0 \text{ if } f[i,j] > \text{threshold} \end{array}$$

Tool: A term used loosely to define something mounted on the end of the robot arm; for example, a hand, a simple gripper, or an arc welding torch.

Transducer.   A device for converting energy from one form to another.

Transfer vector: A transfer table used to communicate between two or more programs. The table is fixed in relationship with the program for which it is the transfer vector. The transfer vector provides communication linkage between that program and any remaining sub-programs.

Trap: An unprogrammed conditional jump to a known location, automatically activated by hardware, with the location from which the jump occurred.

TTY: Teletypewriter.

Variable: A quantity that can assume any of a given set of values.

Vision system: A device that collects data and forms image that can be interpreted by a robot computer to determine the position or to "see" an object.

Word: A set of bits comprising the smallest addressable unit of information in a programmable memory.

Word length: The number of bits in a word.

Write: To deliver data to a medium such as storage.

Zero suppression: The elimination of nonsignificant zeros in a numeral.

Appendix C

A Glossary of Commonly Used Microprocessor Terms In Robotics

Absolute addressing: See *direct addressing.*

Absolute indexed addressing: The effective address is formed by adding the index register (*X* or *Y*) to the second and third byte of the instruction.

Accumulator: A register that holds one of the operands and the result of arithmetic and logic operations that are performed by the central processing unit. Also commonly used to hold data transferred to or from I/O devices.

Accumulator addressing: One-byte instruction operating on the accumulator.

ACIA: Asynchronous Communications Interface Adapter. This is an NMOS LSI device produced by Motorola for interfacing Serial ASCII devices to a microprocessor system.

Address: A number that designates a memory or I/O location.

Address bus: A multiple-bit output bus for transmitting an address from the CPU to the rest of the system.

Algorithm: The sequence of operation that defines the solution to a problem.

Alphanumeric: Pertaining to a character set that contains both letters and numerals and usually other characters.

Courtesy of Feedback, Inc., Berkeley Heights, N.J.

ALU [arithmetic/logic unit]:   The unit of a computing system that performs arithmetic and logic operations.

ASCII code: American Standard Code for Information Interchange. A seven-bit character code without the parity bit, or an eight-bit character code with the parity bit.

Assembler: A program that translates symbolic operation codes into machine language, symbolic addresses to memory addresses, and assigns values to all program symbols. It translates source programs to object programs.

Assembly directive: A mnemonic that modifies the assembler operation but does not produce an object code (e.g., a pseudo-instruction).

Assembly language: A collection of symbolic labels, mnemonics, and data that are to be translated into binary machine codes by the assembler.

Asynchronous: Not occurring at the same time, or not exhibiting a constant repetition rate; irregular.

Base: See *radix*.

BCD: Binary Code Decimal. A means by which decimal numbers are represented as binary values, where integers in the range 0 to 9 are represented by the 4-bit binary codes from 0000 to 1001.

Bidirectional data bus: A data bus in which digital information can be transferred in either direction.

Binary: The base 2 number systems. All numbers are expressed as powers of 2. As a consequence, only two symbols (0 and 1) are required to represent any number.

Bit: The smallest unit of information that can be represented. A bit may be in one of two states, represented by the binary digits 0 and 1.

Block diagram: A diagram in which the essential units of any system are drawn in the form of blocks, and their relationship to each other is indicated by appropriately connected lines.

Branch instruction: An instruction that causes a program jump to a specified address and execution of the instruction at that address. During the execution of the branch instruction, the central processor replaces the contents of the program counter with the specified address.

Breakpoint: Pertaining to a type of instruction, instruction digit, or other condition used to interrupt or stop a computer at a particular place in a program. A place in a program where such an interruption occurs or can be made to occur.

Buffer: A noninverting digital circuit element that may be used to handle a large fan-out or to invert input and output levels. A storage device used to compensate for a difference in rate of flow of data, or time of occurrence of events, when transmitting data from one device to another.

Byte: A sequence of eight adjacent binary digits operated on as a unit.

Call: A special type of jump in which the central procesor is logically required to "remember" the contents of the program counter at the time that the jump occurs. This allows the processor later to resume execution of the main program, when it is finished with the last instruction of the subroutine.

Cascade: An arrangement of two or more similar circuits in which the output of one circuit provides the input of the next.

Clock: A device or a part of a device that generates all the timing pulses for the coordination of a digital system. System clocks usually generate two or more clock phases. Each phase is a separate square-wave pulse train output.

Coding: The process of preparing a program from the flow chart defining an algorithm.

Compiler: A language translator that converts individual source statements into multiple machine instructions. A compiler translates the entire program before it is executed.

Complement: To reverse all binary bit values (ones become zeros, zeros become ones).

Conditional: In a computer, subject to the result of a comparison made during computation.

Conditional breakpoint instruction: A conditional jump instruction that causes a computer to stop if a specified switch is set. The routine then may be allowed to proceed as coded, or a jump may be forced.

Conditional jump: Also called *conditional transfer of control.* An instruction to a computer that will cause the proper one of two (or more) addresses to be used in obtaining the next instruction, depending on some property of one or more numerical expressions or other conditions.

Contact bounce: The uncontrolled making and breaking of a contact when the switch or relay contacts are closed. An important problem in digital circuits, where bounces can act as clock pulses.

CPU [central processing unit]: The unit of a computing system that controls the interpretation and execution of instructions; includes the ALU.

Data bus: A multiline, parallel path over which digital data are transferred from any of several destinations. Only one transfer of information can take place at any one time. While such transfer is taking place, all other sources that are tied to the bus must be disabled.

Debounced: Refers to a switch or relay that no longer exhibits contact bounce.

Debug: Detect, locate, and correct problems in a program or hardware.

Decoder/driver: A code conversion device that can also have sufficient voltage or current output to drive an external device such as a display or a lamp monitor.

Demultiplexer: A digital device that directs information from a single input to one of several outputs. Information for output-channel selection usually is presented to the device in binary weighted form and is decoded internally. The device also acts as a single-pole multiposition switch that passes digital information in a direction opposite to that of a multiplexer.

Destination: Register, memory location, or I/O device that can be used to receive data during instruction execution.

Device select pulse: A software-generated positive or negative clock pulse from a computer that is used to strobe the operation of one or more I/O devices, including individual integrated circuit chips.

Direct addressing: The second and third bytes of the instruction contain the address of the operand to be used.

DMA [direct memory access]: Suspension of processor operation to allow peripheral units external to the CPU to exercise control of memory for both READ and WRITE without altering the internal state of the processor.

Dynamic RAM: A random access memory that uses a capacitive element for storing a data bit. They require REFRESH.

EBCDIC: Extended Binary Coded Decimal Interchange Code, a digital code primarily used by IBM. It closely resembles the half-ASCII code.

Edge: The transition from logic 0 to logic 1, or from logic 1 to logic 0, in a clock pulse.

Editor: A program used for preparing and modifying a source program or other file by addition, deletion, or change.

Effective address: The actual address of the desired location in memory, usually derived by some form of calculation.

Expansion: The process of inserting a sequence of operations represented by a macro name when the marco name is referenced in a program.

Fall time: The time required for an output voltage of a digital circuit to change from a logic 1 to a logic 0 state.

Fan-out: The number of parallel loads within a given logic family that can be driven from one output mode of a logic circuit.

Fetch: One of the two functional parts of an instruction cycle. The collective actions of acquiring a memory address and then an instruction or data byte from memory.

Field: An area of an instruction mnemonic.

FIFO [first in, first out]: The term applies to the sequence of entering data into and retrieving data from data storage. The first data entered are the first data obtainable with FIFO.

File: A collection of data records treated as a single unit.

Flag: A status bit which indicates that a certain condition has arisen during the course of arithmetic or logical manipulations or data transmission between a pair of digital electronic devices. Some flags may be tested and thus be used for determining subsequent actions.

Flag register: A register consisting of the flag flip-flops.

Flow chart: A symbolic representation of the algorithm required to solve a problem.

Frequency: The number of recurrences of a periodic phenomenon in a unit of time. Electrical frequency is specified as so many cycles per second, or hertz.

Full duplex: A data transmission mode that provides simultaneous and independent transmission and reception.

Half-ASCII: A 64-character ASCII code that contains the code words for numeric digits, alphabetic characters, and symbols, but not keyboard operations.

Half-duplex: A data transmission mode that provides both transmission and reception, but not simultaneously.

Handshake: Interactive communication between two system components, such as between the CPU and a peripheral; often required to prevent loss of data.

Hardware: Physical equipment; mechanical, electrical, or electronic devices.

Hexadecimal: A number system based upon the radix 16, in which the decimal numbers 0 through 9 and the letters A through F represent the 16 distinct states in the code.

High address byte: The eight most significant bits in the 16-bit memory address word. Abbreviated H or HI.

HMOS: A trademark for the high-density N-channel, silicon gate semiconductor.

IC (integrated circuit): (1) A combination of interconnected circuit elements inseparably associated on or within a continuous substrate. (2) Any electronic device in which both active and passive elements are contained in a single package. In digital electronics, the term chiefly applies to circuits containing semiconductor elements.

Immediate addressing: The operand is the second byte of the instruction, rather than its address.

Implied addressing: A one-byte instruction that stipulates an operation internal to the processor. Does *not* require any additional operand.

Increment: To increase the value of a binary word. Typically, to increase the value by 1.

Indexed address: An indexed address is a memory address formed by adding immediate data included with the instruction to the contents of some register or memory location.

Indexed indirect addressing: The second byte of the instruction is added to the contents of the X index register, discarding the carry, to form a zero-page effective address.

Indirect absolute addressing: The second and third bytes of the instruction contain the address for the first of two bytes in memory that contain the effective address.

Indirect address: An address used with an instruction that indicates a memory location or a register that in turn contains the actual address of an operand. The indirect address may be included with the instruction, contained in a register (register indirect address), or contained in a memory location (memory directed indirect address).

Indirect indexed addressing: The second byte of this instruction is a zero-page address. The contents of this zero-page address are added to the Y index register to form the lower 3 bits of the effective address. Then the carry (if any) is added to the contents of the next zero-page address to form the higher 8 bits of the effective address.

Instruction: A statement that specifies an operation and the values or locations of its operands.

Instruction code: A unique binary number that encodes an operation that a computer can perform.

Instruction cycle: A successive group of machine cycles, as few as one or as many as seven, that together perform a single microprocessor instruction within the microprocessor chip.

Instruction decoder: A decoder within a CPU that decodes the instruction code into a series of actions that the computer performs.

Instruction register: The register that contains the instruction code.

Interfacing: The joining of members of a group (such as people, instruments, etc.) in such a way that they are able to function in a compatible and coordinated fashion.

Interpreter: A language translator that converts individual source statements into multiple machine instructions by translating and executing each statement as it is encountered. Cannot be used to generate object code.

Interrupt: In a computer, a break in the normal flow of a system or routine such that the flow can be resumed from that point at a later time. The source of the interrupt may be internal or external.

I/O device: Input/output device; any digital device, including a single integrated circuit chip, that transmits data on strobe pulses to a computer or receives data or strobe pulses from a computer.

Jump: (1) To cause the next instruction to be selected from a specified storage location in a computer. (2) A deviation from the normal sequence of execution of instructions in a computer.

Label: One or more characters that serve to define an item of data or the location of an instruction or subroutine. A character is one symbol of a set of elementary symbols, such as those corresponding to typewriter keys.

Latch: A simple logic storage element. A feedback loop used in a symmetrical digital circuit, such as a flip-flop, to retain a state.

Leading edge: The transition of a pulse that occurs first.

LED [light-emitting diode]:   A pn junction that emits light when biased in the forward direction.

Level triggered: The state of the clock input, being either logic 0 or logic 1, carries out a transfer of information or completes an action.

LIFO [last in, first out]:   The latest data entered are the first data obtainable from a LIFO stack or memory section.

Listing: An assembler output containing a listing of program mnemonics, the machine code produced, and diagnostics, if any.

Logic: (1) The science dealing with the basic principles and applications of truth tables, switching, gating, etc. (2) See *logical design*. (3) Also called symbolic logic. A mathematical approach to the solution of complex situations by the use of symbols to define basic concepts. The three basic logic symbols are AND, OR, and NOT. When used in Boolean algebra, these symbols are somewhat analogous to addition and multiplication. (4) In computers and information-processing networks, the systematic method that governs the operations performed on information, usually with each step influencing the one that follows. (5) The systematic plan that defines the interactions of signals in the design of a system for automatic data processing.

Logical decision: The ability of a computer to make a choice between two alternatives; basically, the ability to answer yes or no to certain fundamental questions concerning equality and relative magnitude.

Logical design: The synthesizing of a network of logical elements to perform a specified function. In digital electronics, these logical elements are digital electronic devices, such as gates, flip-flops, decoders, counters, etc.

Logical element: In a computer or data-processing system, the smallest building blocks that operators can represent in an appropriate system of symbolic logic. Typical logical elements are the AND gate and the flip-flop.

Loop: A sequence of instructions that is repeated until a conditional exit situation is met.

Low address byte: The 8 least significant bits in the 16-bit memory address word. Abbreviated L or LO.

LSB [least significant bit]: The digit with the lowest weighting in a binary number.

LSI [large-scale integration]: Integrated circuits that perform complex functions. Such chips usually contain 100 to 2000 gates.

Machine code: A binary code that a computer decodes to execute a specific function.

Machine cycle: A subdivision of an instruction cycle during which time a related group of actions occurs within the microprocessor chip. In the 8080 microprocessor, there exist nine different machine cycles. All instructions are combinations of one or more of these machine cycles.

Macro assembler: An assembler routine capable of assembling programs that contain and reference macro instructions.

Macro instruction: A symbol that is used to represent a specified sequence of source instructions.

Magnetic core: A type of computer storage that employs a core of magnetic material with wires threaded through it. The core can be magnetized to represent a binary 1 or 0.

Magnetic disk: A flat circular plate with a magnetic surface on which data can be stored by selective magnetization of portions of the flat surface.

Magnetic drum: A storage device consisting of a rapidly rotating cylinder, the surface of which can be easily magnetized and which will retain the data. Information is stored in the form of magnetized spots (or no spots) on the drum surface.

Magnetic tape: A storage system based on the use of magnetic spots (bits) on metal or coated-plastic tape. The spots are arranged so that the desired code is read out as the tape travels past the read–write head.

Masking: A process that uses a bit pattern to select bits from a data byte for use in a subsequent operation.

Memory: Any device that can store logic 1 and logic 0 bits in such a manner that a single bit or group of bits can be accessed and retrieved.

Memory address: A 16-bit binary number that specifies the precise memory location of a memory word among the 65,536 different possible memory locations.

Memory cell: A single storage element of memory, capable of storing one bit of digital information.

Microcomputer: A computer system based on a microprocessor and containing all the memory and interface hardware necessary to perform calculations and specified information transformations.

Microprocessor: A central processing unit fabricated as one integrated circuit.

Microprogram: A computer program written in the most basic instructions or subcommands that can be executed by the computer. Frequently, it is stored in a read-only memory.

Mnemonic: Symbols representing machine instructions designed to allow easy identification of the functions represented.

Module: The module of a counter is the number of distinct states the counter goes through before repeating. A four-bit binary counter has a module of 16; a decade counter has a module of 10; and a divide-by-7 counter has a module of 7. In a variable module counter, n can be any value within a range of values.

Monitor: Software or hardware that observes, supervises, controls, or verifies system operation.

Monostable multivibrator: Also called one-shot multivibrator, single-shot multivibrator, or start–stop multivibrator. A circuit having only one stable state, from which it can be triggered to change the state, but only for a predetermined interval, after which it returns to the original state.

MSB [most significant bit]: The digit with the highest weighting in a binary number.

MSI [medium-scale integration]: Integrated circuits that perform simple, self-contained logic systems, such as counters and flip-flops.

Multiplexer: A digital device that can select one of a number of inputs and pass the logic level of that input on to the output. Information for input-channel selection usually is presented to the device in binary weighted form and decoded internally. The device acts as a single-pole multiposition switch that passes digital information in one direction only.

Negative edge: The transition from logic 1 to logic 0 in a clock pulse.

Negative-edge triggered: Transfer of information occurs on the negative edge of the clock pulse.

Negative logic: A form of logic in which the more positive voltage level represents logic 0 and the more negative level represents logic 1.

Nesting: A sequential calling of subroutines without returning to the main program.

Nibble: A sequence of four adjacent bits, or half a byte, is a nibble. A hexadecimal or BCD digit can be represented in a nibble.

Nonoverlapping two-phase clock: A two-phase clock in which the clock pulses of the individual phases do not overlap.

Nonvolatile memory: A semiconductor memory device in which the stored digital data are not lost when the power is removed.

Octal: A number system based upon the radix 8, in which the decimal numbers 0 through 7 represent the eight distinct states.

One-byte instruction: An instruction that consists of eight contiguous bits occupying one successive location.

Open-collector output: An output from an integrated circuit device in which the final "pull-up" resistor in the output transistor for the device is missing and must be provided by the user before the circuit is completed.

Operand: Data that are, or will be, operated upon by an arithmetic/logic instruction; usually identified by the address portion of an instruction, explicitly or implicitly.

Operation: Moving or manipulating data in the CPU or between the CPU and peripherals.

Page: A page consists of all the locations that can be addressed by 8 bits (a total of 256 locations) starting at 0 and going through 255. The address within a page is determined by the lower 8 bits the address, and the page number (0 through 255) is determined by the higher 8 bits of a 16-bit address.

Parity: A method of checking the accuracy of binary numbers. If even parity is used, the sum of all the 1's in a number and its corresponding parity bit is always even. If odd parity is used, the sum of all the 1's and the parity bit is always odd.

Partitioning: The process of assigning specified portions of a system responsibility for performing specified functions.

PC: See *program counter.*

Peripheral: A device or subsystem external to the CPU that provides additional system capabilities.

PIA: Peripheral Interface Adaptor (MOS technology's MPS 6520).

Polling: Periodic interrogation of each of the devices that share a communications line to determine whether it requires servicing. The multiplexer or control station sends a poll that has the effect of asking the selected device, "Do you have anything to transmit?"

Pop: Retrieving data from a stack.

Port: A device or network through which data may be transferred or where device or network variables may be observed or measured.

Positive edge: The transition from logic 0 to logic 1 in a clock pulse.

Positive-edge triggered: Transfer of information occurs on the positive edge of the clock pulse.

Positive logic: A form of logic in which the more positive voltage level represents logic 1 and the more negative level represents logic 0.

Priority: A preferential rating. Pertains to operations that are given preference over other system operations.

Processor: Shorthand word for microprocessor.

Program: A group of instructions that causes the computer to perform a specified function.

Program counter: A register containing the address of the next instruction to be executed. It is automatically incremented each time program instructions are executed.

Program label: A symbol that is used to represent a memory address.

PROM [programmable read-only memory]:   A read-only memory that is field programmable by ther user.

Propagation delay: A measure of the time required for a logic signal to travel through a logic device or a series of logic devices. It occurs as the result of four types of circuit delays:—storage, rise, fall, and turn-on-delay—and is the time between when the input signal crosses the threshold-voltage point and when the responding voltage at the output crosses the same voltage point.

Pseudo-instruction: A mnemonic that modifies the assembler operation but does not produce an object code.

Pull-up resistor: A resistor connected to the positive supply voltage to the output collector of open-collector logic. Also used occasionally with mechanical switches to ensure the voltage of one or more switch positions.

Pulse width: Also called *pulse length*. The time interval between the points at which the instantaneous value on the leading and trailing edges bears a specified relationship to the peak pulse amplitude.

Push: Putting data into a stack.

Radix: Also called the *base*. The total number of distinct marks or symbols used in a numbering system. For example, since the decimal numbering system uses 10 symbols, the radix is 10. In the binary numbering system, the radix is 2, because there are only two marks or symbols (0 and 1). In the octal numbering system, the radix is 8, and in the hexadecimal numbering system, the radix is 16.

RAM [random access memory]:   A semiconductor memory into which logic 0 and logic 1 states can be written (stored) and then read out again (retrieved).

Read: In semiconductors, to transmit data from a semiconductor memory to some other digital electronic device. The term, "read" also applies to computers and other types of memory devices.

Refresh: The process by which dynamic RAM cells recharge the capacitive node to maintain the stored information. The charged nodes discharge due to leakage currents, and without refresh the stored data would be lost. This process must recur every so many microseconds. During refresh, the RAM cannot be accessed.

Refresh logic: The logic required to generate all the refresh signals and timing.

Register: A hardware element used to temporarily store data.

Relative address: A relative address is a memory address formed by adding the immediate data included with the instruction to the contents of the program counter or some other register.

Reset: A computer system input that initializes and sets up certain registers in the CPU and throughout the computer system. One of the initializations is to load a specific address into the program counter. The two bytes of information in that and the succeeding address are the starting address for the system program (for the MOS technology processors).

Return: A special type of jump in which the central processor resumes execution of the main program at the contents of the program counter at the time that the jump occurred.

Ripple counter: A binary counting system in which flip-flops are connected in series.

Rise time: The time required for an output voltage of a digital circuit to change from a logic 0 to a logic 1 state.

ROM [read-only memory]: A semiconductor memory from which digital data can be repeatedly read out, but cannot be written into, as is the case for a RAM.

Routine: A group of instructions that causes the computer to perform a specified function, e.g., a program.

Scratch pad: The term applies to memory that is used temporarily by the CPU to store intermediate results.

Semiconductor memory: A digital electronic memory device in which 1's and 0's are stored that is a product of semiconductor manufacturing.

Seven-segment display: An electronic display that contains seven lines or segments spatially arranged in such a manner that the digits 0 through 9 can be represented through the selective lighting of certain segments to form the digit.

Shift register: A digital storage circuit in which information is shifted from one flip-flop of a chain to the adjacent flip-flop upon application of each clock pulse. Data may be shifted several places to the right or left, depending on additional gating and the number of clock pulses applied to the register. Depending on the number of positions shifted, the rightmost characters are lost in a right shift, and the leftmost characters are lost in a left shift.

Simulator: A program that represents the functioning of one computer system utilizing another computer system.

Software: The means by which any defined procedure is specified for computer execution.

Source: Register, memory location, or I/O device that can be used to supply data for use by an instruction.

Source program: A group of statements conforming to the syntax requirements of a language processor.

Split data bus: Two data buses, one for incoming communications and one for outgoing communications. An 8-bit data bus in split data bus system takes 16 lines.

Stack: A specified section of sequential memory locations used as a LIFO (last in, first out) file. The last element entered is the first one available for output. A stack is used to store program data, subroutine return addresses, processor status, etc.

Stack pointer [SP]: A register that contains the address of the system read/write memory used as a stack. It is automatically incremented or decremented as instructions perform operations with the stack.

Statement: An instruction in source language.

Static RAM: A random access memory that uses a flip-flop for storing a binary data bit. Does not require refresh.

String: A series of values.

Subroutine: A routine that causes the execution of a specified function and that also provides for transfer of control back to the calling routine upon completion of the function.

Symbol: Any character string used to represent a label, mnemonic, or data constant.

Symbolic address: Also called *floating address*. In digital computer programming, a label chosen in a routine to identify a particular word, function, or other information that is independent of the location of the information within the routine.

Symbolic code: A code by which programs are expressed in source language; that is, storage locations and machine operations are referred to by symbolic names and addresses that do not depend upon their hardware-determined names and addresses.

Symbolic coding: In digital computer programming, any coding system using symbolic rather than actual computer addresses.

Synchronous: Operation of a switching network by a clock pulse generator. All circuits in the network switch simultaneously, and all actions take place synchronously with the clock.

Syntax error: An occurrence in the source program of a label, expression, or condition that does not meet the format requirements of the assembler program.

Table: A data structure used to contain sequences of instructions, addresses, or data constants.

Three-byte instruction: An instruction that consists of 24 contiguous bits occupying three successive memory locations.

Three-state device or **Tristate device:** A semiconductor logic device in which there are three possible output states: (1) a logic 0 state, (2) a logic 1 state, or (3) a state in which the output is, in effect, disconnected from the rest of the circuit and has no influence upon it.

Trailing edge: The transition of a pulse that occurs last, such as the high-to-low transition of a positive clock pulse.

Transition: The instance of changing from one state to a second state.

Truth table: A tabulation that shows the relation of all output logic levels of a digital circuit to all possible combinations of input logic levels in such a way as to characterize the circuit functions completely.

Two-byte instruction: An instruction that consists of 16 contiguous bits occupying two successive memory locations.

Two-phase clock: A two-output timing device that provides two continuous series of timing pulses from the second series always following a single clock pulse from the first series. Depending on the type of two-phase clock, the pulses in the first and second series may or may not overlap each other. Usually identified as phase 1 and phase 2.

Unconditional: Not subject to conditions external to the specific computer instruction.

Unconditional call: A call instruction that is unconditional.

Unconditional jump: A computer instruction that interrupts the normal process of obtaining the instructions in an ordered sequence and specifies the address from which the next instruction must be taken.

Unconditional return: A return instruction that is unconditional.

VLSI [very large scale integration]:   Monolithic digital integrated circuit chips with a typical complexity of 2000 or more gates or gate-equivalent circuits.

Volatile memory: A semiconductor memory device in which the stored digital data are lost when the power is removed.

Weighting: Most counters in the 7400 series of integrated circuit chips are weighted counters; that is, we can assign a weighted value to each of the flip-flop outputs in the counter. By summing the product of the logic state times the weighting value for each of the flip-flops, we can compute the counter state. For example, the weighting factors for a 4-bit binary counter are D = weight of 8, C = weight of 4, B = weight of 2, and A = weight of 1. The binary output, $DCBA = 1101_2$ from a 4-bit binary counter would therefore be 13.

Wired-or circuit: A circuit consisting of two or more semiconductor devices with open collector outputs in which the outputs are wired together. The output from the circuit is at a logic 0 if device A *or* device B *or* device C *or* . . . is at a logic 0 state.

Word: The maximum number of binary digits that can be stored in a single addressable memory location of a given computer system.

Write: In semiconductors and other types of memory devices, to transmit data into a memory device from some other digital electronic device. To WRITE is to STORE.

Zero-page: The lowest 256 address locations in memory. Where the highest 8 bits of address are always 0's and the lower 8 bits identify any location from 0 to 255. Therefore, only a single byte is needed to address a location in zero-page.

Zero-page addressing: The second byte of the instruction contains a zero-page address.

Zero-page indexed addressing: The second byte of the instruction is added to the index register (X or Y) to form a zero-page effective address. The carry (if any) is dropped.

Appendix D
A Glossary of Servo Terms Used in Robotics

Accuracy: The ability of a control system, or measuring device, to reproduce faithfully true values of the quantity being measured, such as temperature, light, displacement, speed, etc.

Amplitude: Of an oscillatory signal, the measure of how large it is, e.g., the difference between positive and negative peaks of the signal, although other measures may be specified.

Attenuation: The reduction in the magnitude of a signal as it passes through a system. Opposite of gain or amplification.

Automatic controller: A device or system that measures a quantity, compares this with a desired value, and if necessary initiates a correcting action, e.g., thermostatic controller for central heating.

Automatic error correction: A technique employed within a controlled system to detect errors and eventually correct them.

Automation: When a machine, process, or system, is operated by mechanical, or electronic devices, and requires minimum human participation during the controlled operation, this is known as automation.

Backlash: In a mechanical system, undesired freedom of movement between interacting parts, such as gearwheels.

Courtesy of Feedback, Inc., Berkeley Heights, N.J.

Band: The range of values over which effective control can be applied to a process or system.

Closed loop: Describes a control operation in which an output quantity is compared with an input representing its desired value, the resultant error signal being used to correct the output quantity.

Control potentiometer: Usually a continuously variable resistor, connected to a voltage source, used as the input demand setting, or connected to the output to produce measured value potentials.

Damping: Any agency, or its effect, that reduces or prevents a system's tendency to oscillate.

Dead band: The inactive range, or neutral zone, over which no control is effected by demands at the input.

Derivative action (rate action): A control operation in which the corrective action is related to the rate at which a quantity is deviating from the desired value.

Desired value: The value of an output quantity that is desired, or a signal representing that value, applied as input signal to a control system.

Error: The difference between the desired value and the actual or measured value of an output quantity.

Feedback: The application of a signal, derived from a point (usually the output) of a control system, to an earlier point in the system. Negative feedback means feedback that opposes the effect on the output of disturbances to the system. Positive feedback has the opposite effect and may cause oscillation.

Following error: If the input shaft velocity of a servo system is constant, the output shaft will rotate at the same speed, but lag behind with an angle just sufficient to provide the error signal necessary to maintain the drive. This error is known as the following error.

Frequency response: If a system is subjected to a sine-wave input whose amplitude is constant, but whose frequency varies, the amplitude of the output and its phase relative to the input sine wave will vary. This variation related to frequency is called the frequency response.

Gain (amplification): The increase in signal amplitude as it passes through a system.

Hunting: This occurs when a control system's output continuously searches for a final value. Very rapid hunting is usually termed oscillation.

Hysteresis: The inherent lag in a system or process response when the input is cycled.

Integrator: A device that responds to an input signal not immediately, but with an output that grows in proprotion to the input and the time it is applied.

Linearity: If a change of the input function to a device or system is reproduced in exact proportion at the output, the device or system is said to be linear.

Measured value: The actual quantity attained at the output of the system, whether it be angular displacement, velocity, torque, liquid level, pressure, etc.

Natural frequency: The frequency at which a system would oscillate with no damping applied if a disturbance were injected.

On–off control [bang-bang or two-step control]: The ouput from the system or device is either fully on or completely off.

Open loop: A system in which there is no automatic error correcting, due to the absence of a feedback signal.

Overshoot: The amount by which the change in measured value at the output exceeds the desired changes after a sudden change at the input.

Position sensor: A device, such as a potentiometer, that measures displacement and converts the information into a signal that can be used by a control system.

Proportional control: A control system in which the correcting signal is directly proportional to the error (deviation).

Servomechanism: A closed-loop system for controlling the position or motion of, or force on, a mechanical output element.

Set point: Desired value.

Transducer: A device that converts one signal or physical quantity to another, e.g., a thermostat, a photoelectric cell, a potentiometer.

Transfer function: A mathematical expression for the ratio $\frac{\text{output}}{\text{input}}$ for electrical and mechanical devices.

Transient response: The behavior of the output of a system when the input signal undergoes a sudden change.

Velocity feedback: In a speed-controlled servomechanism, a signal proportional to the velocity of the output shaft is used as feedback and improves the speed regulation during system disturbances. In a position-control system this velocity feedback is used for damping the system's response.

Appendix E

Servomechanism
Conversion Factors

Courtesy of the Singer-Kearfott Division.

	Multiply:	By:	To Obtain:
ANGULAR MEASURE	degrees	17.45	mils
	degrees	60	minutes
	degrees	1.745×10^{-2}	radians
	mils	5.730×10^{-2}	degrees
	mils	3.438	minutes
	mils	1.000×10^{-3}	radians
	minutes	1.667×10^{-2}	degrees
	minutes	0.2909	mils
	minutes	2.909×10^{-4}	radians
	radians	57.30	degrees
	radians	1.000×10^{3}	mils
	radians	3.438×10^{3}	minutes
ANGULAR VELOCITY	deg/sec	1.745×10^{-2}	rad/sec
	deg/sec	0.1667	rpm
	deg/sec	2.778×10^{-3}	rps
	rad/sec	57.30	deg/sec
	rad/sec	9.549	rpm
	rad/sec	0.1592	rps
	rpm	6.0	deg/sec
	rpm	0.1047	rad/sec
	rpm	1.667×10^{-2}	rps
	rps	360	deg/sec
	rps	6.283	rad/sec
	rps	60	rpm
	ARMY MIL	$1/6400$	revolutions
DAMPING	$\dfrac{\text{ft-lb}}{\text{rad/sec}}$	20.11	$\dfrac{\text{oz-in}}{\text{rpm}}$
	$\dfrac{\text{oz-in}}{\text{rpm}}$	4.974×10^{-2}	$\dfrac{\text{ft-lb}}{\text{rad/sec}}$
	$\dfrac{\text{oz-in}}{\text{rpm}}$	6.75×10^{-2}	newton-m/rad/sec
DENSITY	g/cm³	10^{3}	kg/m³
	lb/ft³	16.018	kg/m³
DISTANCE	cm	10^{-2}	meters
	in	2.5400×10^{-2}	meters
	ft	0.30480	meters
	yd	0.91440	meters
	km	10^{3}	meters
	miles	1609.4	meters
ENERGY	ergs	10^{-7}	joules
	kwhr	3.6×10^{6}	joules
	calories	4.182	joules
	ft-lb	1.356	joules
	Btu	1055	joules

	Multiply:	By:	To Obtain:
FORCE AND WEIGHT	dynes	10^{-5}	newtons
	poundals	0.13826	newtons
	lb (force)	4.4482	newtons
INERTIA	g-cm²	10^{-7}	kg-m²
	g-cm²	5.468×10^{-3}	oz-in²
	g-cm²	7.372×10^{-8}	slug-ft²
	oz-in²	1.829×10^{2}	g-cm²
	oz-in²	1.348×10^{-5}	slug-ft²
	slug-ft²	1.357×10^{7}	g-cm²
	(lb-ft-sec²)	7.419×10^{4}	oz-in²
	slug-ft²	1.357	kg-m²
	lb-in²	2.925×10^{-4}	kg-m²
	oz-in²	1.829×10^{-5}	kg-m²
MASS	g	10^{-3}	kilograms
	slug	14.594	kilograms
POWER	ergs/sec	10^{-7}	watts
	cal/sec	4.182	watts
	Btu/hr	0.2930	watts
	joules/sec	1.00	watts
	horsepower	746	watts
	ft-lb/sec	1.356	watts
PRESSURE	dynes/cm²	10^{-1}	newton/m²
	psi	6.895×10^{3}	newton/m²
	atmospheres	1.013×10^{5}	newton/m²
	cm Hg	1333	newton/m²
TORQUE	ft-lb	1.383×10^{4}	g-cm
	ft-lb	192	oz-in
	g-cm	7.235×10^{-5}	ft-lb
	g-cm	1.389×10^{-2}	oz-in
	oz-in	5.208×10^{-3}	ft-lb
	oz-in	72.01	g-cm
	oz-in	7.0612×10^{-3}	newton-m (joules)
TORQUE ERROR	$\dfrac{\text{oz-in}}{\text{minute}}$	0.0558	$\dfrac{\text{lb-ft}}{\text{rad}}$
	$\dfrac{\text{lb-ft}}{\text{rad}}$	17.9	$\dfrac{\text{oz-in}}{\text{minute}}$
VELOCITY	ft/sec	0.30480	m/sec
	miles/hr	0.44704	m/sec
	knots	1.152	miles/hr

Appendix F

A Listing of Robot Manufacturers in the United States

Acco Industries, Inc.
Industrial Lifters Division
Route 37 North
P.O. Box 298
Salem, Ill. 62881

Accumatic Machinery Corp.
3537 Hill Avenue
Toledo, Ohio 43607

Acrobe Positioning Systems, Inc.
3219 Doolittle Drive
Northbrook, Ill. 60062

Admiral Equipment Co.
Sub. of the Upjohn Company
305 West North Street
Akron, Ohio 44303

Advanced Robotics Corp.
Newark Ohio Industrial Park
Route 79—Bldg. 8
Hebron, Ohio 43025

American Robot Corp.
201 Miller Street, Suite 7
P.O. Box 10767
Winston-Salem, N.C. 27103

American Technologies
Division of American Can Co.
1900 Pollitt Drive
Fair Lawn, N.J. 07410

Anorad
111 Oser Avenue
Hauppague, New York 11788

Note: This listing will vary from year to year. Addresses may also change. So do the manufacturers themselves.

Armax Robotics, Inc.
Sub. of Maxco, Inc.
38700 Grand River Avenue
Farmington Hills, Mich. 48108

ASEA, Inc.
Industrial Robot Division
1176 East Big Beaver Road
Troy, Mich. 48084

Automatix, Inc.
217 Middlesex Turnpike
Burlington, Mass. 01803

Automaton Corp.
Affiliate of Marathon Industries
23996 Freeway Park Drive
Farmington Hills, Mich. 48024

Bendix Corp.
Robotics Division
21238 Bridge Street
Southfield, Mich. 48034

Binks Manufacturing Co.
9201 West Belmont Avenue
Franklin Park, Ill. 60131

Cincinnati Milacron, Inc.
Industrial Robot Division
215 S. West Street
Lebanon, Ohio 45036

Comet Welding Systems
900 Nicholas Boulevard
Elk Grove Village, Ill. 60007

Control Automation, Inc.
Clarksville and Everett Road
P.O. Box 2304
Princeton, N.J. 08540

Copperweld Robotics, Inc.
1401 East 14 Mile Road
Troy, Mich. 48084

Cybotech Corp.
P.O. Box 88514
Indianapolis, Ind. 46208

Cyclomatic Industries, Inc.
Sub. of Palco Industries, Inc.
7520 Convoy Court
San Diego, Calif. 92111

DeVilbiss Company, The
300 Phillips Avenue
P.O. Box 913
Toledo, Ohio 43692

Dynamco, Inc.
Air Controls Division
2639 Manana Drive
Dallas, Tex. 75220

Elicon
273 Viking Avenue
Brea, Calif. 92621

ESAB North America, Inc.
Robotic Welding Division
P.O. Box 2286
1941 Heath Parkway
Fort Collins, Colo. 80522

Fared Robot Systems, Inc.
3860 Revere Street, Suite D
P.O. Box 39268
Denver, Colo. 80239

Feedback, Inc.
620 Springfield Avenue
Berkeley Heights, N.J. 07922

Gallaher Enterprises, Inc.
2110 Cloverdale Avenue, Suite 2–B
P.O. Box 10244
Winston-Salem, N.C. 27108

GCA/PaR Systems
3460 Lexington Avenue, North
St. Paul, Minn. 55112

General Electric Co.
Automation Systems
1285 Boston Avenue
Bridgeport, Conn. 06602

General Numeric Corp.
390 Kent Avenue
Elk Grove Village, Ill. 60007

Graco Robotics, Inc.
12898 Westmore Avenue
Livonia, Mich. 48150

Hirata Corp. of America
8900 Keystone Crossing
Indianapolis, Ind. 46240

Hitachi America, Ltd.
Industrial Components
8 Pearl Court
Allendale, N.J. 07401

Hobart Brothers Co.
Hobart Square
Troy, Ohio 45373

Hodges Robotics International Corp.
3710 North Grand River Avenue
Lansing, Mich. 48906

IBM Corp.
System Products Division
2000 N.W. 51st Street
P.O. Box 1328
Boca Raton, Fla. 33432

Ikegal America Corp.
2246 No. Palmer Drive
Suite 108
Schaumberg, Ill. 60195

Industrial Automates, Inc.
6123 W. Mitchell Street
Milwaukee, Wis. 53214

International Robomation/Intelligence
2281 Las Palmas Drive
Carlsbad, Calif. 92008

I.S.I. Manufacturing, Inc.
31915 Groesbeck Highway
Fraser, Mich. 48026

C. Itoh & Co., America, Inc.
21415 Civic Center Drive
Suite 111
Southfield, Mich. 48076

Kuka Welding Systems & Robot Corp.
24031 Research Drive
Farmington Hills, Mich. 48018

Kulicke & Soffa Industries, Inc.
507 Prudential Road
Horsham, Pa. 19044

Lamson Corp.
P.O. Box 4857
Syracuse, N.Y. 13221

Lynch Machinery Division
2300 Crystal Street
P.O. Box 2477
Anderson, Ind. 46018

Mack Corp.
3695 East Industrial Drive
Flagstaff, Ariz. 86001

Manca, Inc.
Link Drive
Rockleigh, N.J. 07647

Microbot, Inc.
453–H Ravendale Drive
Mountain View, Calif. 94043

Mobot Corp.
980 Buenos Avenue, Suite A
San Diego, Calif. 92110

Nordson Corp.
Robotics Division
555 Jackson Street
P.O. Box 151
Amherst, Ohio 44001

Pentel of America Ltd.
2715 Columbia Street
Torrance, Calif. 90503

Pickomatic Systems, Inc.
37950 Commerce
Sterling Heights, Mich. 48077

Positech Corp.
Rush Lake Road
Laurens, Iowa 50554

Prab Robots, Inc.
5944 East Kilgore Road
Kalamazoo, Mich. 49003

Precision Robots, Inc.
6 Carmel Circle
Lexington, Mass. 02173

Reeves Robotics, Inc.
Box 5
Issaquah, Wash. 98027

Reis Machines
1426 Davis Road
Elgin, Ill. 60120

Rexnord, Inc.
Hydraulic Products Division
P.O. Box 383
Milwaukee, Wis. 53201

Rhino Robots, Inc.
3085 State Street
Champaign, Ill. 61820

Rob-Con, Ltd.
12001 Globe Road
Livonia, Mich. 48150

Robogate Systems, Inc.
A Comau Affiliate
750 Stephenson Highway, Suite 3000
Troy, Mich. 48084

Robomation Corp.
P.O. Box 652
Madison Heights, Mich. 48071

Robotics, Inc.
R.D. 3, Route 9
Ballston Spa, N.Y. 12020

Robotic Sciences International, Inc.
136 West 157 Street
Gardena, Calif. 90428

Schrader Bellows
Division of Scovill, Inc.
200 West Exchange Street
Akron, Ohio 44309

Seiko Instruments, U.S.A., Inc.
2990 West Lomita Boulevard
Torrance, Calif. 90505

Shin Meiwa
c/o Nichimen Co., Inc.
8 N. Michigan Avenue
Chicago, Ill. 60602

Sigma Sales, Inc.
65–5 Serrano Avenue, Ste. C
Anaheim Hills, Calif. 92807

Sormel, SA
c/o Ceeris International, Inc.
1055 Thomas Jefferson Street, N.W.
Suite 414
Washington, D.C. 20007

Sterline Detroit Co.
261 East Goldengate Avenue
Detroit, Mich. 48203

Textron, Inc.
Bridgeport Machines Division
500 Lindley Street
Bridgeport, Conn. 06606

Thermwood Corp.
P.O. Box 436
Dale, Ind. 47523

Tokico America, Inc.
3555 West Lomita Boulevard, Suite E
Torrance, Calif. 90505

Transcom, Inc.
8824 Twinbrook Road
P.O. Box D
Mentor, Ohio 44060

Unimation, Inc.
Division of Westinghouse Electric Corp.
Shelter Rock Lane
Danbury, Conn. 06810

United States Robots
1000 Conshohocken Road
Conshohocken, Pa. 19428

United Technologies Corp.
Automotive Group
5200 Auto Club Drive
Dearborn, Mich. 48126

Wear Control Technology, Inc.
189–15 Station Road
Flushing, N.Y. 11358

Westinghouse Electric Corp.
Industry Automation Division
400 Media Drive
Pittsburgh, Pa. 15205

Yaskawa Electric America, Inc.
305 Era Drive
Northbrook, Ill. 60062

Appendix G

General Bibliography on Robots and Robotics

Manipulation

A. Bejczy, "Sensors, Controls, and Man-Machine Interface for Advanced Teleoperation," *Science*, vol. 208, no. 4450, June 20, 1980, pp. 1327–35.

M. Brady, ed., *Robot Motion, Planning and Control*, MIT Press, Cambridge, Mass., 1982.

T. L. Brooks, "Superman: A System for Supervisory Manipulation and the Study of Human/Computer Interactions," ONR Contract No. N00014–77–C–0256, Master's thesis, Man-Machine Systems Laboratory, Massachusetts Institute of Technology, Cambridge, Mass., 1979.

R. L. Paul, *Robot Manipulators, Mathematics, Programming, and Control*, MIT Press, Cambridge, Mass., 1981.

J. K. Salisbury, "Articulated Hands: Force Control and Kinematic Issues," *The International Journal of Robotics Research*, vol. 1, no. 1 Spring 1982.

Courtesy of SRI International, Menlo Park, Calif.

General and Industrial Robots

R. G. Abraham, R. J. Stewart, and L. Y. Shum, "State of the Art in Adaptable-Programmable Assembly Systems," Westinghouse R&D Center (IFS Publications, Ltd., 35–39 High Street, Kempston, Bedford MK42 7BT, England).

J. S. Albus, *Brains, Behavior, and Robotics*, BYTE Books, Peterborough, N.H., 1981, 352 pages.

John J. Allan, *A Survey of Industrial Robots*, 1st ed., Productivity International, Inc., 1980.

A. Baar and E. A. Feigenbaum, *The Handbook of Artificial Intelligence*, Stanford University, Stanford, Calif. (HeurishTech Press), 1981, 409 pages.

Wayne Hwa-Wei Chen, *The Year of the Robot*, Dilithium Press, 1981.

"Computerized Robots, The State of the Art," Integrated Computer Systems, Inc., ed., 3304 Pico Blvd., P. O. Box 5339, Santa Monica, Calif., 90405 (Sept. 1980).

Joseph Duffy, *Analysis of the Mechanisms of Robot Manipulators*, John Wiley & Sons, Inc., New York, 1980.

J. Engelberger, *Robotics in Practice, Management and Applications of Industrial Robots*, AMACOM, 135 West 50th St., New York, New York, 1981.

General Dynamics Corp., "ICAM Robotics Application Guide," Technical Report AFWAL–TR:80–4042, Vol. II, Materials Laboratory, Air Force Wright Aeronautical Laboratories, Air Force Systems Command, Wright-Patterson Air Force Base, Ohio 45433 (1980).

R. M. Glorioso and F. C. Colon Osorio, *Engineering Intelligent Systems*, Digital Press, Bedford, Mass., 1980, 472 pages.

R. Goldman, "Recent Work with the AL System," *Proceedings, 5th International Joint Conference on Artificial Intelligence—1977*, Massachusetts Institute of Technology, Cambridge, Mass. (August 22–25, 1975), pp. 733–735.

James A. Gupton, *Microcomputers for External Control Devices*, Dilithium Press, Beaverton, Oregon, 1980.

V. Daniel Hunt, *Industrial Robotics Handbook*, Industrial Press Inc., New York, 1983.

Industrial Robots, A Survey of Foreign and Domestic U.S. Patents, 1982, Available from the U.S. Department of Commerce, National Technical Information Service, 5295 Port Royal Road, Springfield, Virginia, 22161 at $95 per copy.

David F. Iver and Rober W. Bolz, *Robotics Sourcebook and Dictionary*, Industrial Press, Inc., New York, 1983.

J. C. Latombe and A. Lux, "Intellegence Artificielle et Robotique Industrielle," *Le Nouvel Automatisme*, no. 6 (May 1979), pp. 37–44 and no. 7, (June–July 1979), pp. 21–29.

G. Lundstrom, *Industrial Robots–Gripper Review*, IFS Publications, England, 1977.

P. McCorduck, *Machines Who Think*, W. H. Freeman and Company, San Francisco, 1979.

D. Nitzan et al., "Machine Intelligence Research Applied to Industrial Automation," Tenth Report, National Science Foundation Grant DAR78–27128, SRI International, Menlo Park, Calif., Nov. 1980.

Richard P. Paul, *Robot Manipulators*, MIT Press, Cambridge, Mass., 1981.

B. Raphael, *The Thinking Computer: Mind Inside Matter*, W. H. Freeman and Co., San Francisco, 1976, 322 pages.

"The Road to the Automatic Factory, 1970–1981," *Manufacturing Engineering* (Jan. 1982), pp. 209–251.

Robots 7: Conference Proceedings and Supplement (1983), SME, Publication Sales, One SME Drive, Dearborn, Mich. 48128. Published yearly.

K. Sadamoto, ed., *Robots in the Japanese Economy*, Tokyo, 1981, 256 pages.

G. Sullivan, *Rise of the Robots*, Dodd, Mead and Co., New York, 1971, 114 pages.

Ken Susnjara, *The Manager's Guide to Industrial Robots*, Technical Database[R] Corp., Conroe, Texas, 1982.

W. R. Tanner, ed., *Industrial Robots, Second Edition, Vol. I, Fundamentals*, Society of Manufacturing Engineers, Dearborn, Mich., 1981, 282 pages.

W. R. Tanner, ed., *Industrial Robots, Second Edition, Vol. II, Applications*, Society of Manufacturing Engineers, Dearborn, Mich. 1981, 287 pages..

D. Tesar, "Trip Report, Visits to Major Research Centers in Robotics in Europe and Russia," Center for Intelligent Machines and Robotics, Department of Mechanical Engineering, University of Florida, Gainesville, 1981.

Martin B. Weinstein, *Android Design*, Hayden Book Co., Rochelle Park, N.J., 1981.

P. Weintraub, "Raising the Robot's I.Q.," *Discover*, vol. 2, no. 6 (June 1981), pp. 76–80.

Welding Institute, Developments in Mechanized, Automated and Robotic Welding (conference), 1981.

P. H. Winston, *Artificial Intelligence*, Addison-Wesley Publishing Co., Reading, Mass., 440 pages.

P. H. Winston, and R. H. Brown, *Artificial Intelligence: An MIT Perspective. Vol. I: Expert Problem Solving, Natural Language Understanding, Intelligent Computer Coaches, Representation and Learning*, MIT Press, Cambridge, Mass., 1979, 492 pages.

P. H. Winston, and R. H. Brown, *Artificial Intelligence: An MIT Perspective. Vol. II: Understanding Vision, Manipulation, Computer Design, Symbol Manipulation*, MIT Press, Cambridge, Mass., 1979, 486 pages.

Walking Robots

D. E. Orin, "Supervisory Control of a Multilegged Robot," *International Journal of Robotics Research*, vol. 1, no. 1 (Spring 1982).

Sensing

H. H. Baker and T. O. Binford, "Depth from Edge and Intensity Based Stereo," *Proceedings, 7th International Joint Conference on Artificial Intelligence,* Vancouver, British Columbia, Aug. 1981.

T. O. Binford, "Survey of Model-Based Image Analysis Systems," *International Journal of Robotics Research,* vol. 1, no. 1 (Spring 1982).

G. G. Dodd, and L. Rossol, eds., *Computer Vision and Sensor-Based Robots,* Plenum Publishing Corp., New York, 1979.

J. Free, "New Eyes for Computers: Chips That See," *Popular Science* (Jan. 1982), pp. 61–63.

L. D. Harmon, "Touch-Sensing Technology: A Review," Technical Report MSR80–03, Society of Manufacturing Engineers, One SME Drive, P. O. Box 930, Dearborn, Mich., 48128 (1980).

S. T. Kowel, "Beyond Optical Imagers: The DEFT Camera," *Optical Spectra* (July 1980), pp. 52–56.

G. Mamon et al., "Pulsed GaAs Laser Terrain Profiler," *Applied Optics,* vol. 18, no. 6 (Mar. 15, 1978), pp. 868–889.

D. Nitzan, "Assessment of Robotic Sensors," *Proceedings, NSF Robotics Research Workshop,* Newport, R.I., April 15–17, 1980 (reprints available from Robotics Dept., SRI International, 333 Ravenswood Ave., Menlo Park, Calif. 94025).

D. Nitzan, A. E. Brain, and R. O. Duda, "The Measurement and Use of Registered Reflectance and Range Data in Scene Analysis," *Proceedings of the IEEE,* vol. 65, no. 2 (Feb. 1977), pp. 206–220.

Photonics Spectra, vol. 16, issue 1 (Jan. 1982), pp. 59–60.

P. Stucki, ed., *Advances in Digital Image Processing, Theory, Application, Implementation,* Plenum Publishing Corp., New York, 1979.

E. R. Teja, "Voice Input and Output," *Electronic Design News* (Nov. 20, 1979).

J. M. Tenenbaum et al., "Prospects for Industrial Vision," A. I. Center Technical Note 175, SRI International, Menlo Park, California, Nov. 1978.

Computers

H. Foley, "A Preliminary Survey of Artificial Intelligence Machines," *SIGART Newsletter,* no. 72 (July 1980), pp. 21–28.

E. R. Hnatek, "Semiconductor Memory Update: DRAMS," *Computer Design,* vol. 21, no. 1 (Jan. 1982), pp. 109–120.

M. Marshall, "LISP Computers Go Commercial," *Electronics* (Nov. 20, 1980), pp. 89–90.

D. Smith and D. Miller, "Computing at the Speed of Light," *New Scientist* (Feb. 21, 1980), pp. 554–556.

Appendix H

Organizations, Magazines,
Journals, Catalogs, and Other
Periodicals Involved in Robots
and Robotics, Conferences,
and Courses

Organizations

U.S.A

American Association for Artificial Intelligence (AAAI), 445 Burgess Drive, Menlo Park, California 94025. Phone (415) 328–3123.

American Federation of Information Processing Societies (AFIPS), 11815 North Lynn Street, Suite 800, Arlington, Virginia 22209. Phone: (703) 558–3600.

American Institute of Industrial Engineers, Inc., 25 Technology Park/Atlanta, Norcross, Georgia 30092.

Association for Computing Machinery (ACM), 11 West 42nd Street, Third Floor, New York, New York 10036. Phone: (212) 869–7440.

International Institute for Robotics, Box 210708, Dallas, Texas 75211.

Robot Institute of America, P. O. Box 930, Dearborn, Michigan 48128. Phone:
   (313) 271–0778.

Robotics International (RI/SME), P.O. Box 930, One SME Drive, Dearborn,
   Michigan 48128. Phone: (313) 271–1500.

Japan

Japan Industrial Robot Association, c/o Kikaishinko Bldg., 3–5–8, Shibakoen,
   Minato-ku, Tokyo, Japan.

Europe

British Robot Association, 35–39 High Street, Kempston Bedford MK42 7BT,
   United Kingdom.

Danish Robot Association, c/o Technological Institute, Division of Industrial
   Automation, Gregersensvej, 2630 Taastrup, Denmark.

French Industrial Robot Association, 91. rue Falguiere, 75015 Paris, France.

Italian Society for Industrial Robots, c/o Etas Kompass, Via Mantegna, 6,
   Milano, Italy.

Swedish Industrial Robot Association, Storgatan 19, Box 5506, S–114 85
   Stockholm, Sweden.

Others

Singapore Robotic Association, 5, Portsdown Road, Off Ayer Rajah Road,
   Singapore 0513, Republic of Singapore.

Magazines, Journals, Catalogs, and Other Periodicals

1983 Robotics Industry Directory, $35.00, Robotics Industry Directory, P. O. Box 725,
   La Canada, California 91011. Published yearly.

The International Journal of Robotics Research, M. Brady and R. L. Paul, eds., quarterly, $45.00/year, ISSN 0278–3649, MIT Press Journals, 28 Carleton Street, Cam-
   bridge, Massachusetts 02142.

Robotics Age, monthly, $15.00/year, Robotics Age, Inc., 174 Concord Street, Peterborough, New Hampshire 03458.

Robotics Today, quarterly, $21.00/year, Robotics International of Society of Manufacturing Engineers, One SME Drive, P. O. Box 930, Dearborn, Michigan 48128.

The Industrial Robot, quarterly, $84.00/year, IFS Publications Ltd., 35–39 High Street, Kempston, Bedford MK42 7BT, England.

Assembly Automation, 5 issues/year, International Fluidic Services, IFS Publications Ltd., 35–39 High Street, Kempston, Bedford MK42 7BT, England.

Sensor Review, quarterly, $84.00/year, IFS Publications Ltd., 35–39 High Street, Kempston, Bedford MK42 7BT, England.

Industrial Robots International, monthly newsletter, $12.00/copy, Technical Insights, Inc., 158 Linwood Plaza, P. O. Box 1304, Fort Lee, New Jersey 07024.

Proceedings, International Conferences on Industrial Robot Technology, first through eleventh, IFS Publications Ltd., 35–39 High Street, Kempston, Bedford MK42 7BT, England.

Robots News International, International Fluidic Services, IFS Publications Ltd., 35–39 High Street, Kempston, Bedford MK42 7BT, England.

Robotics Conferences *

Robots N—(SME) Annual Conference (fall)
Autofact—(SME) Annual Conference on the automatic factory. Robot sessions are included.
Robots in Manufacturing—Local clinics focused on different applications. SME sponsored.
Nth International Conference on Industrial Robots (Annual)

Robotics Courses *

"Robots in Manufacturing" (2 days, $395) Dr. William Tanner
   Sponsored by SME, covers applications, human relations, justification ROI, robots available, etc.
"Computerized Robots" (4 days $895) ICS Course 375
   ICS, Box 5339, Santa Monica, Calif. 90405. Covers control equations, sensors, computer vision, mechanics, robot survey, group technology, work cell, etc.

---

* Courtesy of Dr. John W. Hill, SRI International and Microbot, Inc.

"Introduction to Robots and Computer Vision," CS227, Dr. Tom Binford, 3 units

"Fundamentals of AI," CS223, Dr. Genesereth, 3 units

    Stanford Institutional TV Network

    Nonregistered Students $162/unit

    Honors Coop Program $296/unit

Appendix I

U.S. Air Force Integrated Computer-Aided Manufacturing Program

Robots and computers are bedfellows on the integrated computer-aided manufacturing (ICAM) program because both can do certain manufacturing tasks faster, more effeciently, at less cost, and with greater quality assurance. While the Air Force does not propose the indiscriminate use of robotics in aerospace companies, neither will it overlook the advantages of using them.

Robots used in industrial applications are typically placed in labor-intensive positions performing vastly repeated single functions. However, due to the nature of aerospace manufacturing, where typically less than 10 finished aircraft roll off the production line each month, relatively small batches of parts are involved in assembly. By adapting the computer controls on the industrial robots to more sophisticated software that allows for the storage of a variety of programs, a single robot is capable of conducting several functions such as drilling and routing at the same work cell.

While the robotics effort at General Dynamics has continued, additional Materials Laboratory programs have included robotics. Moreover, new developments in off-line programming, a high-order language, and enhanced sensory perception promise to further improve future robotic productivity.

Courtesy of the U.S. Air Force Aeronautical Systems Division, Wright-Patterson Air Force Base, Ohio.

Appendix J

U.S. Army Artificial Intelligence Program

For the past one and one-half years the Army has been reviewing potential applications and needs for artificial intelligence and robotics (AI/R) that could enhance its operational capabilities. What began as a Headquarters, Department of the Army (HQDA), exploratory request for information about potentials of these rapidly emerging technologies appears to have grown to a general awakening to potentials throughout the Army. It must be realized that there were a very large number of AI/R related activities ongoing within the Army laboratories. For more information, contact the U.S. Army Engineer Topographic Laboratories, Fort Belvoir, Virginia 22060.

Unmanned tanks can be sent into enemy territory, steering themselves around potholes, unaided to reconnoiter using artificial intelligence.

Information on Artificial Intelligence is also available from SRI International, Robotic Department at Menlo Park, California 94025.

Index

A